Jan Didden
publisher/editor

AF880022

Linear Audo - Volume 3

Authors:

Richard S. Burwen

Stan Curtis

Samuel Groner

Patrick K (EUVL)

Richard Lyons

Richard Marsh

Robert Munnig Schmidt

Ovidiu Popa

Steven van Raalte

Ramkumar Ramaswamy

Sigurd Ruschkowski

Scott Wurcer

Stuart Yaniger

Opportunity is missed by most people because it comes dressed in overalls and looks like work.

Thomas Edison

Linear Audio is published by Linear Audio of Hoensbroek, The Netherlands.
Published in April 2012.

Copyright notice

No part of this publication may be reproduced or transmitted in any form or by any means, electronic or mechanical, including photocopying, recording, or any information storage and retrieval system, without permission in writing by the publisher.
Copyrights of the articles and designs in this publication remain with their original authors.

Important notices and disclaimers

Many of the designs, projects, products, methods, instructions or ideas in Linear Audio publications involve potentially lethal voltages. The information is provided with the understanding that the practitioner has sufficient knowledge and experience to safely execute the ideas and practices described in this publication, and to avoid injury and damages to persons and property. Safe and professional practices are assumed. If you are not sure to have this knowledge and experience, you should seek experienced professional advice and assistance.

Neither the Publisher nor the authors, contributors or editors assume any liability for any injury and/or damage to persons or property as a matter of product liability, negligence or otherwise, or from any use or operation of any projects, products, methods or ideas contained in the material in Linear Audio publications and materials.
The designs, projects, products, methods, instructions or ideas in Linear Audio publications may be subject to patents, trademarks or similar deposits, which are all recognized.

The use of general descriptive or registered names, trademarks or similar identifications in this publication does not imply that these are free for general use.

Design, Prepress: Fred Zurel, F&N Eigen Beheer - www.eigenbeheer.nl
Production: F&N Eigen Beheer, Castricum, The Netherlands
ISBN: 9789490929008

www.linearaudio.net

Guest Editorial

Audible tweaks that cannot work

Richard S. Burwen

I am a non-believer in audio tweaks, especially power cords that feed regulated power supplies in any piece of equipment. Scientifically they cannot possibly have any effect on the audio signal. Yet I am convinced it is very possible, even likely, that testing them will demonstrate an audible effect. How is that possible?

The answer is: the room acoustics changed - the test was uncontrolled. In a good listening room, successive sound reflections from the ceiling, floor, walls, and any objects in the room add to produce standing waves at any given audio frequency. Because the many signals travel different distances, their differing phase shifts add and subtract to produce different sound levels at different positions. A sine wave sound from two or more speakers will produce standing waves, even without reflections. If you walk across the room while a speaker plays a pure tone, you will hear the sound become louder and softer as you pass through peaks and valleys of sound pressure. The higher the frequency, the shorter is the wavelength, and the closer together are the peaks and valleys. Because the standing wave pattern is different at each different frequency, a graph of the audio frequency response at a fixed microphone position will show hundreds of peaks and valleys, no matter how flat is the equipment and speaker system producing the sound.

All it takes is a change in position of the listener to hear a different frequency response. Furthermore, a change in position of another person or any object in the room will alter the standing wave pattern, thereby changing the frequency response at any particular listening position.

Your ears are surprisingly sensitive to frequency response. What you hear most is the general trend of the response curve, an average of the peaks and valleys, and 1 dB is noticeable. I found that if the sound is very well balanced, adjusting the response as little as 0.1 dB with tone controls becomes important to a critical ear. The peaks and valleys in frequency response produced by room reverberation may be as small as 1 dB, or as great as 30 dB, when nearly complete cancellation occurs. If an instrument in the orchestra happens to play a note that peaks at the listener, it can stand out like a sore thumb. Moving the listener's head a few inches may reduce its level very noticeably.

My own listening studio is very live and the peaks and valleys in frequency response are big. I made a test by feeding a 5000 Hz sine wave to my 5-speaker system and placing a sound level meter on the coffee table in front of my best seat. Doing the best I could, not to move my body, I slid the sound level meter sideways from one major peak to another. The distance between peaks was only 6.7 inches and there were two minor peaks in between. When I rotated my head about 110 degrees I heard 4 peaks and 4 valleys in the sound level. Moving my body about 6 inches created a small valley where the meter was reading on the coffee table. If I moved a glass of water only 1 inch it produced 1 dB change in level at a major peak. If the meter was measuring 10 dB or so down on the side of peak, all I had to move the glass was about ¼ inch to make a 1 dB change in sound level. The glass is a small reflecting object in a 47 foot long room.

How does the change in sound with position relate to overall listening? Here is a little story.

One of my most favorite recordings and a powerful demonstrator is my own live concert recording of Mahler's Symphony 6, played by the Boston Philharmonic in Jordan Hall at the New England Conservatory, Boston, in 1984. One of the instruments of the orchestra is a sledge hammer! For each of Mahler's 3 blows of fate in the final movement, a muscular guy took a full swing and struck a whopping big anvil on the front of the stage. My two special omnidirectional microphones, pointed at the ceiling to get high frequency sound reflections, captured it well.

This was a 2-channel recording made on a Sony PCMF1 machine, now made into 5 channels using my AUDIO SPLENDOR program which added ambiance in the rear and front channels. It was saved as a 6-channel, 88.2 kHz, 32-bit floating-point, W64 music file. I love this recording so much I had re-mastered it 6 times, each time improving the sound for my own studio. After the 7th re-mastering, I called my wife Barbara down to hear this great symphony and sound. She sat in the center seat and I sat beside her. When the recording played, the violins no longer sounded quite right. I thought about one more re-mastering. After my concert finished and Barbara went upstairs, I embarked on fixing the violins. I played the offending musical passages again, but this time the violins sounded good.

I had been using quite a lot of rear speaker sound to add to front channel sound at the listening sofa and sweeten the violins. When Barbara sat in the center seat, she partially blocked sound from the right rear speaker. I figured out what to do. People damage the sound by absorbing and blocking it. So no more visitors will be allowed.

Well, that conflicts with the purpose of my sound studio. Once again I re-mastered Mahler's 6th, increasing the high frequency reverberation 0.2 dB in all channels, increasing the rear channel gain 0.8 dB, and tweaking the tone control sliders a few tenths of a dB. I spent the last 15 minutes deciding on a 0.1 dB increase in the 3 kHz slider. Without it, the orchestra was more natural. With it, the cymbals were more natural.

As you can see, frequency response is so critical and sensitive to the listening position and reflecting and absorbing objects, it is impractical to make a truly controlled listening test on a tweak. To accurately compare one power chord with another, every person and object in the room would have to remain in the exact same position within about 1/8 wavelength at 20 kHz, only 0.08 inches. This effect is in addition to the Power of Expectation. So the controversy will go on.

Contents

Letters to the Editor: I encourage readers' comments on articles in Linear Audio, and I regularly receive them. I post these on-line at www.linearaudio.net, together with the author's reply.

How Discrete Signal Interpolation Improves Digital-to-Analog Conversion

Richard Lyons

We encounter the process of digital-to-analog conversion every day— not just in our home audio systems, but also in telephone calls (land lines and cell phones), telephone answering machines, CD & DVD players, iPhones, digital television, MP3 players, digital radio, and even talking greeting cards. This is a brief tutorial on how sample rate conversion improves the quality of digital-to-analog conversion.

Ideal Digital-to-Analog Conversion

Have a look at the system shown in **Figure 1**(a). There I show a hardware depiction of a digital-to-analog converter (DAC)—a hardware device having multiple input pins that accept multibit binary *words*. In that figure the variable $x(n)$ represents a sequence of binary words showing their individual binary bits from the least significant bit (LSB) to the most significant bit (MSB). In Figure 1(b) I show a hypothetical time-domain sequence of $x(n)$ amplitude values, which we'll call "samples", where each sample is represented as a single black dot. We refer to the $x(n)$ signal as a discrete, or "digital", sequence. The variable n is referred to as the "time-domain index" of the discrete $x(n)$ input signal.

Critical to its operation, the DAC accepts a periodic-in-time, pulse-like, signal shown as f_{clk} in Figure 1(a). The repetition rate, the frequency, of the f_{clk} signal is the reciprocal of the time period between individual $x(n)$ samples. The f_{clk} signal, synchronized with the binary $x(n)$ input sequence, triggers the DAC to 'clock in' the bits of the current $x(n)$ sample.

Finally, the DAC has a single output pin upon which is riding an analog (what the digital signal processing experts call "continuous") voltage that we'll call $v_{DAC}(t)$. The variable t in $v_{DAC}(t)$ represents time measured in seconds. Given the $x(n)$ input sequence shown in Figure 1(b), we wish to generate the analog $v_{ideal}(t)$ signal shown in Figure 1(c) whose frequency-domain spectrum is depicted in Figure 1(d). In our frequency plots we only show the positive-frequency axis because we assume all signals are real-valued and all spectra are symmetrical and centered at zero Hz.

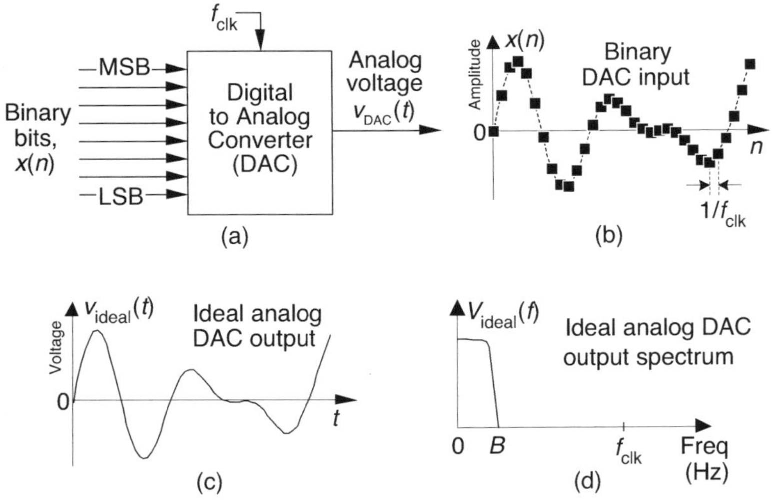

Figure 1

The Nyquist Criterion

In this discussion, if the highest-frequency spectral component of $x(n)$, and therefore $v_{ideal}(t)$, is B Hz, we're assuming that the f_{clk} sample rate is greater than $2B$ Hz. That condition is the famous *Nyquist criterion* stipulating the periodic sampling condition required for error-free sampling (analog-to-digital conversion) of analog signals. As a historical note, the notion of periodic sampling was studied by various engineers, scientists, and mathematicians such as the Russian V. Kotelnikov, the Swedish-born H. Nyquist, the Scottish E. Whittaker, and the Japanese I. Someya[1]. But it was the American mathematician Claude Shannon, acknowledging the work of others, who mathematically formalized the concept of periodic sampling as we know it today, named it in honor of the great American electrical engineer Harry Nyquist, and brought it to the broad attention of the world's communications en-

Claude Shannon

Harry Nyquist

[1] Luke, H. "The Origins of the Sampling Theorem," *IEEE Communications Magazine*, April 1999, pp. 106–109.

gineers[2]. That was in 1948—the birth year of the transistor, marshmallows, and this author.

OK, back to Figure 1. Unfortunately, commercially-available DACs will not produce our desired $v_{ideal}(t)$ output voltage based on the $x(n)$ input sequence. So, given that apparently bad news, let's now think about some hypothetical, and actual, DAC output signals and their spectra.

Some Hypothetical DAC Output Signals

If we could build a DAC whose output voltage was a periodic series of super-narrow (widths measured in picoseconds) analog pulses as shown in **Figure 2**(a), whose amplitudes are equal the to amplitudes of $x(n)$, the spectrum of such analog pulses would be the repetitive pattern shown in Figure 2(b).

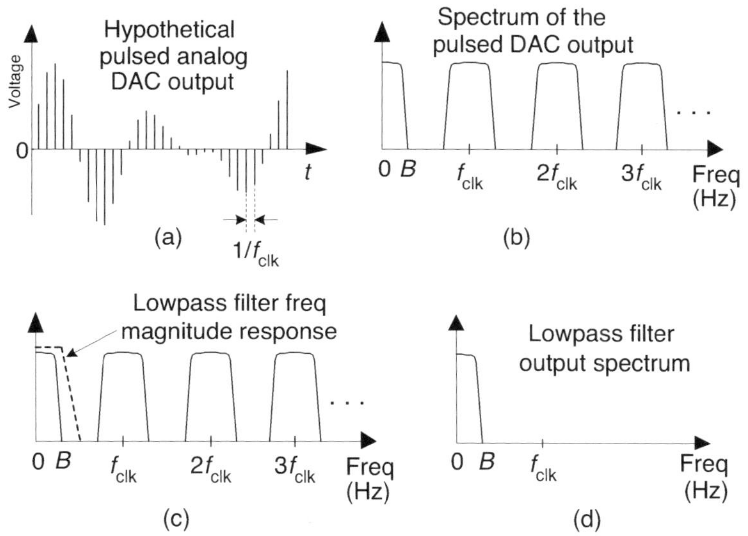

Figure 2

Figure 2 illustrates one of the relationships between the time- and frequency-domain representations of a signal: if a time signal has periodic amplitude variations, such as the periodically-spaced pulses in Figure 2(a), its spectrum will be periodic. If our time signal's super-narrow pulses are separated by $1/f_{clk}$ seconds, the repetitive spectral energy will be separated by f_{clk} Hz as shown in Figure 2(b).

Now if we passed the analog pulsed signal in Figure 2(a) through an analog lowpass filter, whose frequency magnitude response is shown by the dashed lines in Figure 2(c), we'd produce our desired Figure 1(c) $v_{ideal}(t)$ signal having the spectrum shown in Figure 2(d). As straightforward as all of this seems to be, as it turns out, the electronics needed to generate the super-narrow Figure 2(a) pulses is prohibitively expensive, on the order of the cost of a new Harley Davidson Sportster motorcycle. This is far too costly to include in any telephone, music, or television product. Thankfully, DAC man-

[2] Shannon, C. "A Mathematical Theory of Communication," *Bell Sys. Tech. Journal*, Vol. 27, 1948, pp. 379–423, 623–656.

ufacturers have a far more affordable way of generating our desired $v_{ideal}(t)$ signal.

Thinking again about Figure 1, we can view the discrete $x(n)$ sample values in Figure 1(b) as being amplitude samples of our desired $v_{ideal}(t)$ analog signal in Figure 1(c). What we want is an interpolated version of $x(n)$. But not merely two or four or ten new samples between each original $x(n)$ sample, we seek an infinite number of samples between each original $x(n)$ sample. We want so many new samples that our interpolated signal is continuous (analog). It's not traditional interpolation we desire; we wish to perform *interpolation on steroids*. This notion of interpolation is not unusual. Some signal processing experts refer to the DAC process itself as interpolation[3].

Thinking about our idea of interpolation, it would be nice if we could build a DAC whose output voltage was that shown by the solid curve in **Figure 3**(a). As it turns out, that solid curve in Figure 3(a) is not a curve at all –it's a series of straight lines connecting the $x(n)$ sample values shown by the shaded dots.

Let's zoom in and think about the first bold-line segment of that curve as shown in Figure 3(b). To generate the voltage segment between the $x(0)$ and $x(1)$ amplitude values we'd have to electronically evaluate the following expression:

$$v_{1st}(t) = x(0) + [x(0) - x(1)]t, \text{ for } 0 \leq t < 1. \qquad (1)$$

Equation (1) is a first-order (first-order in terms of t) polynomial, and that's why the $v_{1st}(t)$ curve comprising the connected straight lines is called a "first-order hold" waveform. (The process of generat-

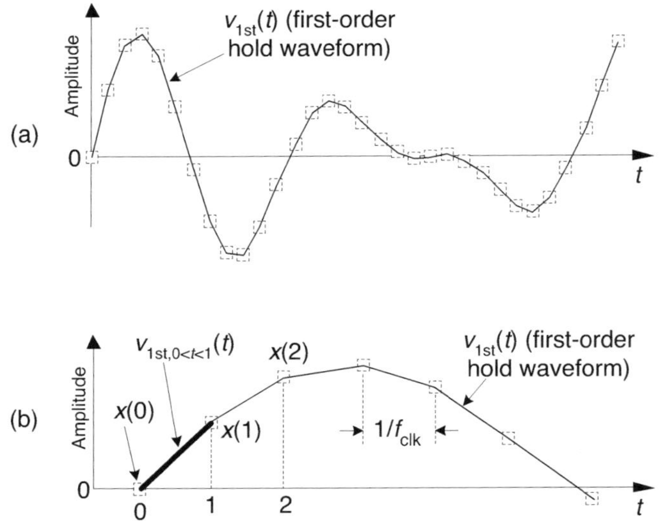

Figure 3

[3] Prandoni, P. and Vetterli, M., *Signal Processing For Communications,* EPFL Press, Lausanne, Switzerland, 2008, pp. 240-247.

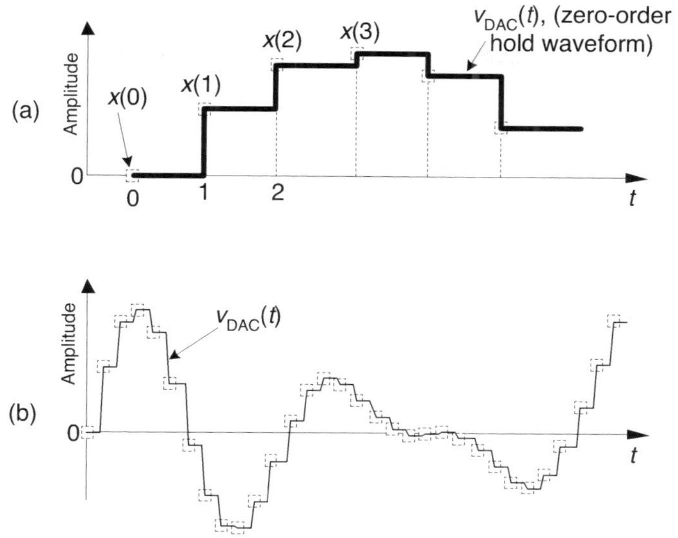

Figure 4

ing the $v_{1st}(t)$ waveform, based on the $x(n)$ samples, is called "curve fitting" by signal processing folk). As with the hypothetical voltage in Figure 2(a), the electronic components needed to generate $v_{1st}(t)$ would be prohibitively expensive for our DAC applications. Let's now review the less-expensive method for generating our desired $v_{ideal}(t)$ signal developed by commercial DAC designers.

Actual DAC Output Signals

Given the discrete $x(n)$ input sequence shown in Figure 1(b), a commercially-available DAC output voltage will be the analog $v_{DAC}(t)$ signal shown in **Figure 4**(a). To generate the $v_{DAC}(t)$ voltage segment between the $x(0)$ and $x(1)$ amplitude values we need merely implement the following expression:

$$v_{DAC}(t) = x(0), \text{ for } 0 \leq t < 1 \tag{2}$$

which is not a function of time t. Likewise, between the $x(1)$ and $x(2)$ amplitude values the $v_{DAC}(t)$ voltage is merely an implementation of:

$$v_{DAC}(t) = x(1), \text{ for } 1 \leq t < 2. \tag{3}$$

Equations (2) and (3) are referred to as "zero-order polynomials" because we could write Eq. (3), for example, as

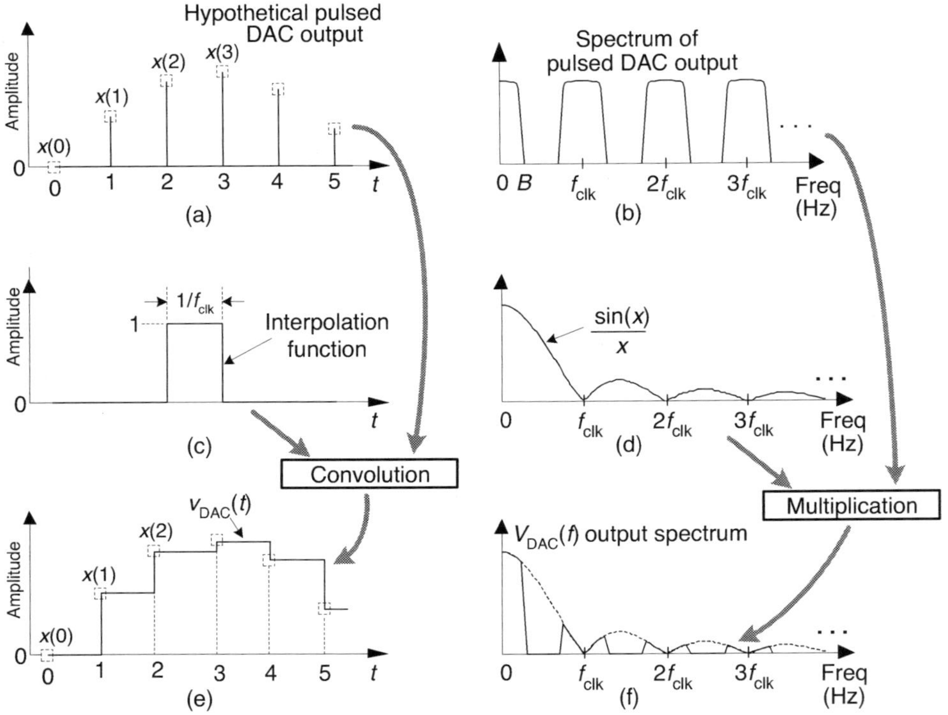

Figure 5

$$v_{DAC}(t) = x(1)t^0, \text{ for } 1 \le t < 2. \tag{4}$$

Notice that the power of variable *t* in Eq. (4) is zero. The $v_{DAC}(t)$ signal is commonly called a "zero-order hold" waveform [3]. Zooming out from Figure 4(a) we see a longer-time interval of the $v_{DAC}(t)$ signal in Figure 4(b).

Now that we know the zero-order hold nature of DAC analog outputs, let's determine the spectral content of such an analog voltage. To do so, we return to the hypothetical analog pulsed DAC output that we considered in Figure 2. We repeat that pulsed signal, and its repetitive spectrum, in **Figures 5**(a) and 5(b). We can think of our DAC's 'stairstep' $v_{DAC}(t)$ output signal as the convolution of the Figure 5(a) pulses with a rectangular, unity-amplitude, interpolation function shown in Figure 5(c) having the sin(x)/x spectrum given in Figure 5(d). Because convolution in the time domain is equivalent to multiplication in the frequency domain, the Figure 5(f) solid-curve spectrum of $v_{DAC}(t)$ is the product of the Figure 5(b) and Figure 5(d) spectra.

So there you have it—the spectrum of a DAC's $v_{DAC}(t)$ output is the repetitive, decreasing-magnitude, spectrum given in Figure 5(f).

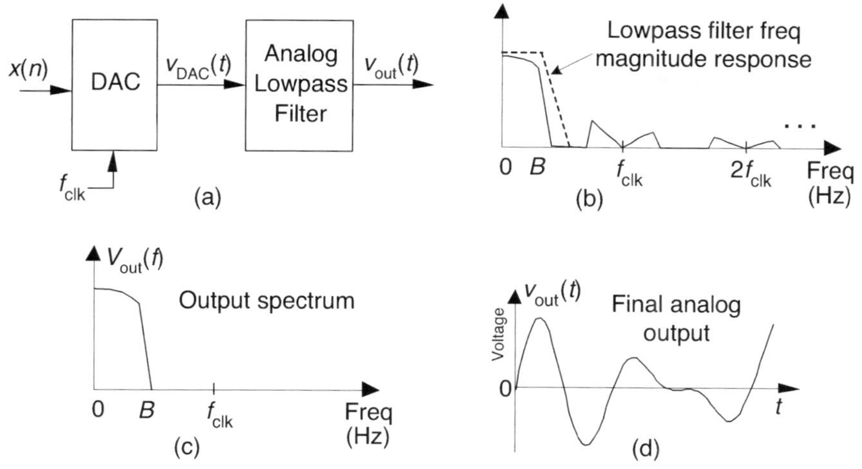

Figure 6

Post-DAC Analog Filtering

Our final task to achieve our ideal Figure 1(c) $v_{ideal}(t)$ voltage, whose spectrum is in Figure 1(d), is to pass the $v_{DAC}(t)$ output signal through an analog lowpass filter as shown in **Figure 6**(a). The lowpass filter's mandatory frequency magnitude response is shown by the dashed lines in Figure 6(b). The spectrum of our resultant $v_{out}(t)$ is given in Figure 6(c), and the $v_{out}(t)$ voltage, in Figure 6(d), is quite similar to our desired $v_{ideal}(t)$ analog signal in Figure 1(c).

OK, there are two things we must consider regarding the DAC/filtering system in Figure 6(a). First, the DAC's non-flat Figure 5(d) sin(x)/x magnitude envelope attenuates the higher frequency spectral components in our original $x(n)$ signal. That is, the drooping nature of a DAC's inherent sin(x)/x frequency characteristic, shown in **Figure 7**(a), attenuates our original signal in the vicinity of B Hz.

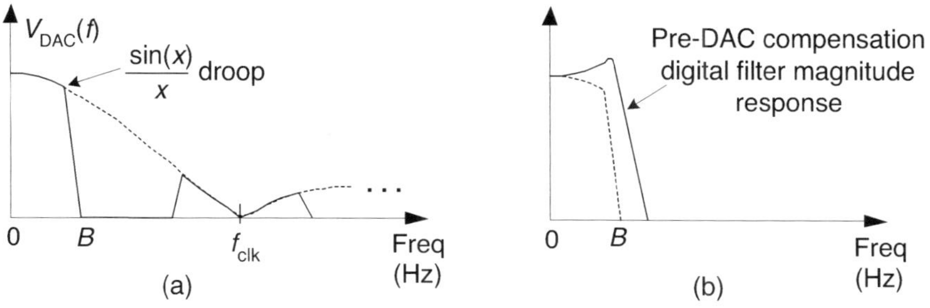

Figure 7

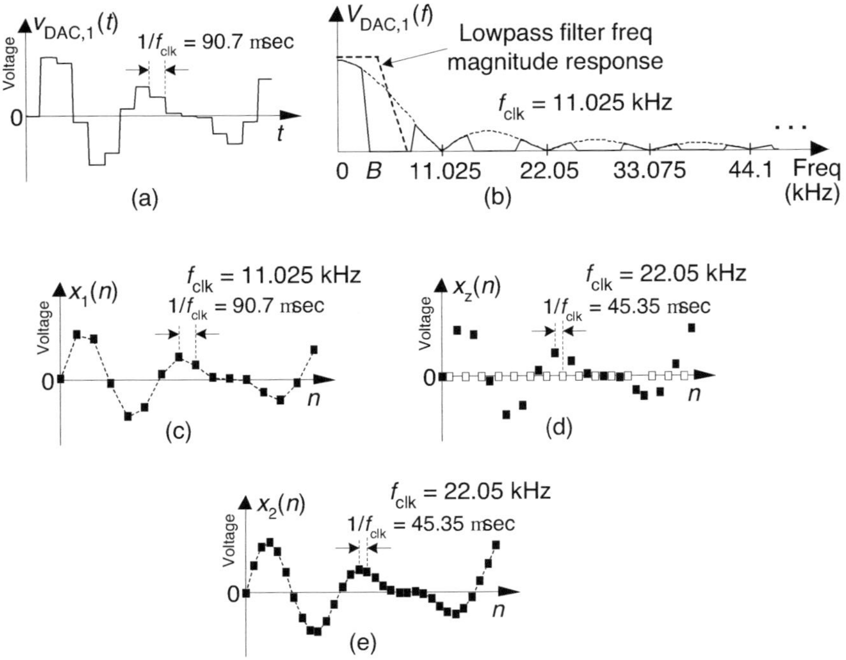

Figure 8

In some DAC applications, if the frequency *B* Hz is small relative to the f_{clk} frequency, perhaps we could simply ignore the sin(*x*)/*x* amplitude droop. However if the sin(*x*)/*x* droop cannot, for some reason, be tolerated, then specialized 'DAC-compensation' digital filtering must be performed in the processing steps prior to applying the *x*(*n*) sequence to the DAC. That is, our signal of interest may need to be applied to a pre-DAC compensation digital filter such as the solid curve in Figure 7(b) where the high-frequency signal components in the vicinity of *B* Hz are amplified. That amplification would then be canceled by the sin(*x*)/*x* droop behavior of our DAC.

The second issue for us to consider regarding the DAC/filtering system in Figure 6(a) is the complexity of the analog lowpass filter. An analog filter having a frequency magnitude response as shown by the dashed lines in Figure 6(b), with such a narrow transition region from the end of its passband to the beginning of its stopband, may be difficult to design. Such a filter may have several active electronic components and be expensive. In addition to the cost issue, analog lowpass filters have the additional problem of exhibiting nonlinear phase responses in the neighborhood of their cutoff frequency at *B* Hz. Fortunately, a multirate digital signal processing operation is available that can reduce both the complexity (cost) and nonlinear phase problems associated with analog lowpass filters used in DAC applications.

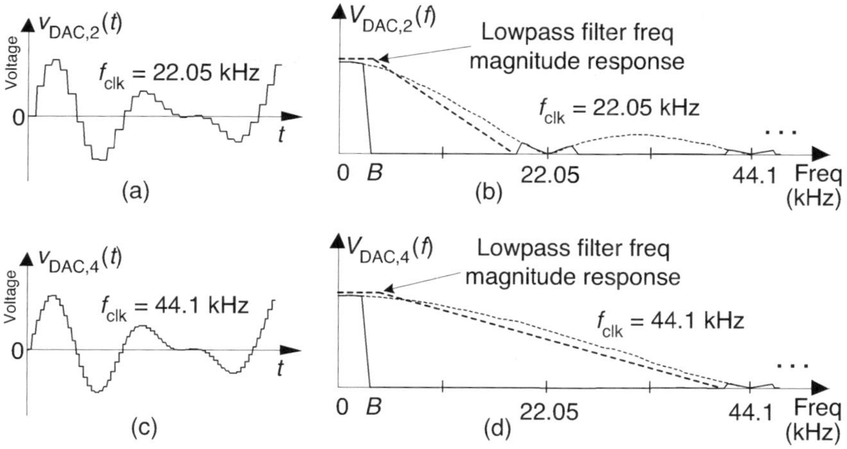

Figure 9

Digital Interpolation to the Rescue

We can drastically simplify our analog lowpass filter design complexity by increasing our $x(n)$ signal's f_{clk} sample rate which is our DAC's f_{clk} clock frequency. That is, we will interpolate our original $x(n)$ signal. To explain this notion of interpolation, consider the $v_{DAC,1}(t)$ analog DAC output signal in **Figure 8**(a). Assuming that the DAC's f_{clk} clock rate is 11.025 kHz, the spectrum of the DAC's output signal is the solid $V_{DAC,1}(f)$ curve in Figure 8(b). At this 11.025 kHz clock rate the analog lowpass filter's frequency magnitude response will be the bold dashed curve in Figure 8(b). We show the DAC's $x_1(n)$ input discrete sequence, whose sample rate is 11.025 kHz, in Figure 8(c).

We implement a 'digital interpolation-by-two' process as follows: First, prior to any DAC processing, we insert a zero-valued sample in between each of the $x_1(n)$ samples to generate the $x_z(n)$ sequence shown in Figure 8(d). (The inserted samples are shown by white dots in Figure 8(d). The full time durations of the $x_1(n)$ and $x_z(n)$ sequences, from their first to their last samples, measured in seconds, are identical.) Next we merely pass the $x_z(n)$ sequence through a digital lowpass filter whose cutoff frequency is slightly greater than *B* Hz. That filter's output, then, will be the 22.05 kHz sample rate interpolated $x_2(n)$ sequence shown in Figure 8(e).

So, in our case, the simplest form of digital interpolation is a two-step process: zero-valued sample insertion followed by digital lowpass filtering. A more thorough discussion of the various forms of digital interpolation can be found in chapter 10 of reference[4].

Applying the interpolated $x_2(n)$ discrete sequence to our DAC generates the $v_{DAC,2}(t)$ voltage shown in **Figure 9**(a). The neat part here is that the spectrum of $v_{DAC,2}(t)$ is that shown in Figure 9(b). The analog lowpass filter, whose frequency magnitude response is the bold dashed curve in Figure 9(b), has a much more gradual (wider) transition region from the end of its passband to the beginning of its

4 Lyons, R., *Understanding Digital Signal Processing* 3rd Ed., Prentice Hall Publishing, Upper Saddle River, NJ, 2010, pp. 507-588.

stopband. This means the $V_{DAC,2}(f)$ lowpass filter is simpler and much less expensive than the $V_{DAC,1}(f)$ lowpass filter.

There are two additional advantages of our factor of two interpolation. first, the undesirable spectral noise centered at 22.05 kHz in Figure 9(b) is smaller in magnitude than the spectral noise centered at 11.025 kHz in Figure 8(b). Second, the digital lowpass filter used in our digital interpolation process can be designed to implement the desirable pre-DAC compensation from Figure 7(b).

We could go one step further and interpolate sequence $x_2(n)$ by another factor of two to generate a discrete $x_4(n)$ sequence having a 44.1 kHz sample rate. Applying that $x_4(n)$ sequence to our DAC will generate the $v_{DAC,4}(t)$ voltage in Figure 9(c) whose spectrum is given in Figure 9(d). The analog low-pass needed for the $v_{DAC,4}(t)$ signal, the bold dashed curve in Figure 9(d), can be quite simple now, perhaps merely a few resistors and capacitors.

Concluding Remarks

We started with a discussion of *ideal* digital-to-analog conversion. Realizing that we cannot perform such an ideal conversion, we considered a few hypothetical digital-to-analog conversion scenarios to help us understand the behavior of commercially-available digital-to-analog converters (DACs). We see that, due to real-world DAC imperfections, analog lowpass filters are needed for proper digital-to-analog conversion. Finally, we showed how digital signal interpolation can be used to drastically reduce the complexity and cost of the analog lowpass filters. (Remember, if we can reduce the cost of our analog filter by 50 cents, and we sell 6 million units, we've saved three million dollars!)

The digital interpolation scheme used to reduce analog circuit complexity is a classic example of how digital signal processing can be used to build lower-cost, more reliable, systems. In addition, such digital interpolation has the benefit of reducing the sin(x)/x droop effects of our DAC, thus improving the quality of the final analog lowpass filter's $v_{out}(t)$ signal.

Design Considerations for a Class A Amplifier Enclosure

Patrick K. (alias EUVL)

In summer 2010, Jan Didden invited me to write an article for Linear Audio. I was somewhat flattered, to find myself in the company of industry names like Pass and Self, but wasn't sure of the subject to write about. Then, folks at the DIY Audio forum (www.diyaudio.com) found out that I have some pretty Class A power amplifier enclosures, as well as access to good quality machining facilities. I agreed to design a Class A power amplifier enclosure and have it manufactured in a limited quantity. Of course, this was just one possible implementation and others may have different requirements. Therefore, I decided to write about the design considerations for such an enclosure; by going through the design process, using this enclosure as an example, the article hopefully provides guidance to those who wish to dimension and design their own.

Estimating total heat dissipation

The most important function of a Class A power amplifier enclosure is without doubt the provision of heat sinking to keep the temperature of the power transistors within their reliable safe operating region. The amount of heat to be dissipated determines the size of the heat sinks, which in turn determines the size and shape of the enclosure. Thus, the first thing one should work out is the total amount of dissipation.

The dissipation of a Class A amplifier itself is relatively easy to determine. It is simply rail to rail voltage times total bias. Take the example of the EUVL F5X, the rails are ±16V, and total bias per channel is 4A [1]. Total dissipation per channel is thus 128W; 256W for 2 channels.

That looks almost too easy, but we need to add a few more heat sources, such as rectifier diodes. To be on the safe side, one should allow at least 1V forward voltage per diode. So for one bridge per channel of F5X, the total diode dissipation is 8W.

With such high current draw, it is not practical to provide sufficient filtering capacitance to flatten the rectified waveform without excessively high ripple currents. So some form of power supply filtering is desirable. The simplest of those is the well-known CRC after the rectifiers. But a resistor produces voltage drop, and heat. For effective 1st order filtering, let's assume an RC corner frequency of say 5Hz. Assume also that each C is 44000uF, R needs to be 0R68, with a voltage drop of 2.7V and a dissipation of 11W per R, or 22W per channel.

But if you are prepared to drop 2.7V, you also have other options. One example is e.g. a MOSFET based follower ´regulator´, or more widely known as capacitor multiplier (**Figure 1**). Let's take a look at the datasheet of the 2SK3497 UHC-MOSFET. The transconductance at 4A is about 10.5S, i.e. the output impedance of the follower regulator is 0.1R. The Vgs at 4A is about 2.45V at room temperature, but dropping to 2.25V at a case temperature of 100°C. Such a high tempco is normally a sign of trouble (thermal run-away) if the device is used in amplifier circuits, as the bias would increase drastically as the device warms up. For the follower regulator application however, it merely means that the dropout voltage decreases by 0.2V with temperature rise. The regulator MOSFET itself gets less heat (dropout voltage x bias current), and the minute rail voltage increase to the amplifier itself is of no significance. The saturation voltage Vds is below 1V at 4A, so that the incoming ripple voltage can be up to 2.2V pk-pk, i.e. equal to the MOSFET Vgs at bias. For the negative rail, one can just use the complementary device, 2SJ618. They are both easily available for reasonable price. On paper at least, the follower regulator provides a better performance than a CRC filter. (But I am also aware that some people swear by passive unregulated power supplies for power amplifier applications).

So, for the example of a stereo enclosure for 2 channels of F5X, the total dissipation amounts to 294W. But we're still not done. Your friendly domestic power company is allowed to fluctuate the mains voltage within a certain margin for load regulation. There might be some tolerances on the transformer voltage under load, etcetera. It is perhaps wise is allow for at least 10% extra, and we end up with something around 330W for 2 channels, or 165W per heat sink.

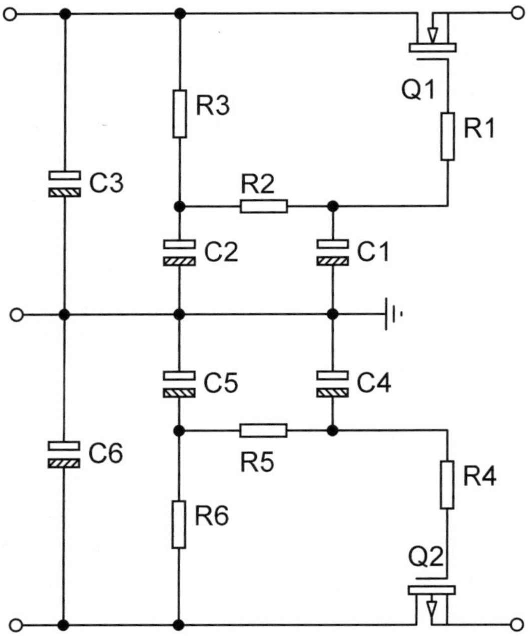

Fig 1 - typical UHC MOSFET follower regulator

Heat sink selection

You may wonder why there is so much discussion about heat sink selection. After all, most heat sink catalogues contain detailed technical sections, and the sinks are normally accompanied by an Rth vs. length curve. Just read the value off, multiplied by the dissipation, and you're done. Right ? Wrong !

The topic is complicated because there is no strict definition of how Rth is applied. Some manufacturers define their Rth at 75°C temperature rise (dT), some at 80°C, some even 90°C, and Rth is not constant with dT.

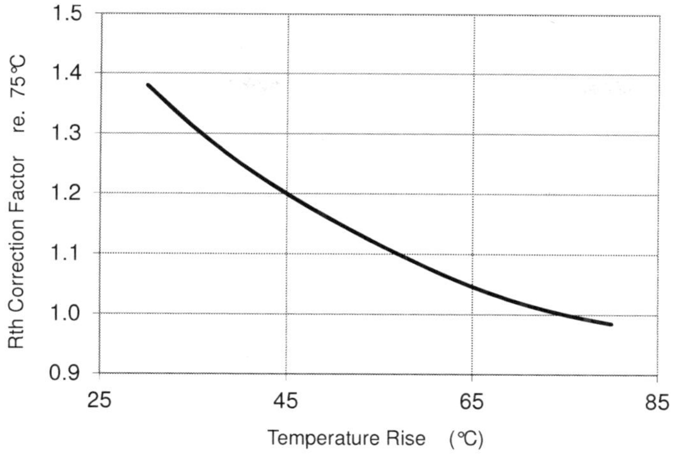

Figure 2a: Rth correction factor versus temperature rise

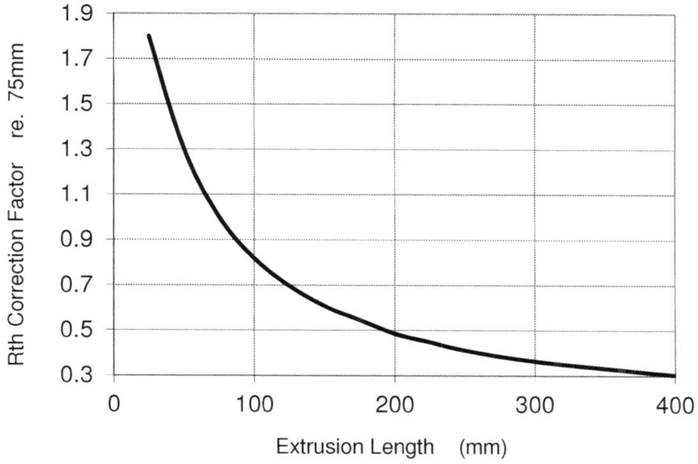

Figure 2b: Rth correction factor versus heatsink length.

Let's start with one particular brand to get a feel for the issues. One well known manufacturer for heat sinks extrusions is Aavid, and they have a quite extensive technical section. For example, they publish a set of correction tables for *their* heat sinks at their webpage [2]. We need to apply 3 correction factors; one for extrusion length, one for dT (starting with an estimate), and one for forced air cooling velocity (ignore if using natural convection, i.e. passive, no fan cooling).

The figures given by Aavid are illustrated in **Figure 2**. **Figure 2a** shows how the Rth of a heatsink referenced to 75°C temperature rise should be corrected for other temperature rise figures. **Figure 2b**, on the other hand, shows how the Rth figure changes with extrusion length. Note however, that these curves may differ significantly between manufacturers, although the general trend is similar. E.g. the correction factor of a Conrad heat sink is listed as 1.33 for dT = 30°C, whereas it is 1.26 for Aavid, a difference of 6%.

Looking at the length correction factor of Aavid, there is still quite some performance to be gained by increasing extrusion length beyond say 150mm. However, this is only a simplified story, for the length correction factor curve not only depends on the length, but also on fin geometry. The reason for that is that at increased length, another dominant factor comes into play namely the resistance to natural convection air flow between the fins. This flow resistance *increases* with increased length, but it also *decreases* rapidly (to the 3rd power as a first approximation) with increased fin pitch.

Figure 3 illustrate this well. While the Rth is still having a useful drop beyond 150mm length, the one with the 10mm pitch almost flattens after 150mm. Both of these sinks have 40mm net fin height, so that the only variable here is really fin pitch.

Thus, as you can see, if you have all the design freedom (e.g. machine your own heat sink or use custom ordered extrusions), you can tailor the heat sink profile to suit your required geometry. E.g., in a tower design using a heat sink of 420mm width x 50mm depth x 500mm height, the optimum fin pitch is about 17mm.
Most of us are not so fortunate to get customised profiles, and are stuck with standard profiles on offer. Many of these, especially those with width > 300mm, have a fin pitch of 10mm. In those cases, there is little point in choosing a heat sink much higher than 150mm. This in turn leads to the commonly seen "flat pack" enclosure designs.

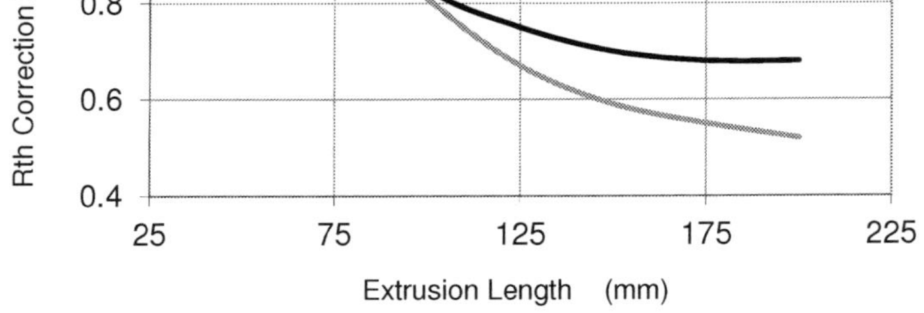

Figure 3: Rth correction factor dependency on fin pitch

As already explained, you cannot just use the Rth figures provided in the catalogue without applying all correction factors. And comparing Rth figures between manufacturers has to be done with great care. Once this is mastered, choosing the right heat sink becomes less of a daunting task.

So now let's look at the other side of the equation, the allowed temperature of the devices to be mounted on the heatsink. Almost all transistor manufacturers quote a maximum working junction temperature of 125 ~ 150 °C. Some manufacturers, while listing these numbers, put in a small note saying that for highest reliability, the design operating temperature should be substantially lower. A detailed description of the failure modes of semiconductors can be found in chapter 2 & 3 of the Toshiba Semiconductor Reliability Handbook [3,4], but even there no single magic number can be found which says make or break. The only figure which can be considered useful is Figure 2.2 in Chapter 3. In general it is wise to allow some 20% margin on the maximum operating power (or 10% voltage / current). This leads to a maximum junction temperature of 100°C as a design goal.

Let's consider again our F5X example. We have a total of 4 power MOSFETs, each dissipating 32W nominal. The maximum junction temperature allowed is 100°C. We also need to consider:

1. the temperature drop from junction to case;
2. the temperature drop across the insulator between the TO247 package and the heat sink.

For 1, most modern TO247 power transistor packages have an Rth of about 0.83 °C/W. At 32W, this equals 27°C. A good insulator such as Keratherm 86/82 has a specific thermal resistance of 35°C.mm^2/W, or 3.5°C for a TO247 with 32W; this is the value for 2. above. The maximum local heat sink temperature at the MOSFET case should thus be not higher than 70°C. As I will show later, the local temperature variation across a heat sink can be as high as 20°C. Thus a calculated *average* heat sink temperature of 60°C would make a good rule of thumb in heat sink dimensioning.

The need for actual thermal measurements

Even with all correction factors applied, the Rth figures still do not give you the entire picture. All manufacturer figures assume a uniformly distributed heat input to the entire back surface of the sink. This is hardly the case in reality. Thus, most heat sink suppliers would always recommend a 3-dimensional thermal simulation, to which most hobbyists do not have access, or thermal measurements using realistic heat sources and real heat sinks under actual working conditions.
Such a thermal measurement has been carried out for the F5X-design using a Conrad MF35-151 heat sink. To recap, the manufacturer's Rth with a 40°C dT correction factor applied is 0.26°C/W. The measurement was carried out by bolting 6 TO247 MOSFETs directly onto the heat sink, and applying a total dissipation of 135W, uniformly distributed between the MOSFETs. The actual steady state temperature on the heat sink is shown in **Figure 4**. The ambient temperature was 22.1°C.

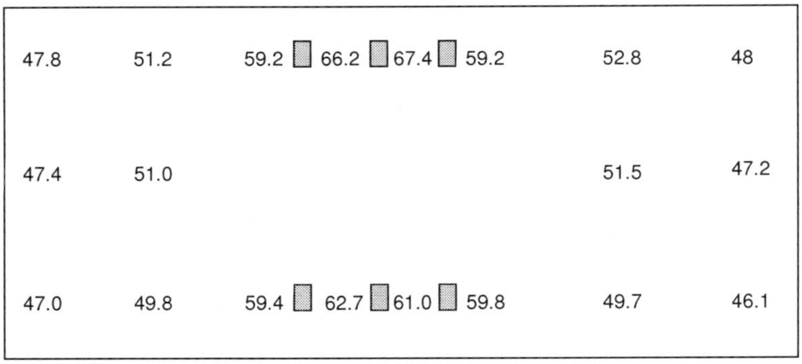

Figure 4: Steady-state temperatures across heatsink with six heat sources as shown (measured)

As could be expected, the top is hotter than the bottom, and temperature rise is highest in the middle. For the average sink temperature of 56.8 degrees, Rth = (56.8 – 22.1) / 135W = 0.26 °C/W. Worst case, highest temperature point 67.4 degrees, Rth = (67.4 – 22.1) / 135W = 0.34 °C/W . (The first number is identical to that given by the manufacturer !!). In this particular setup, if we include the additional dissipation due to rectification and power supply regulation, the MOSFET junction temperature can be calculated as:

22°C + (330W x 0.26°C/W / 2) +10°C + 3°C + 27°C = 105 °C

A bit on the high side, perhaps. The additional surfaces of the front and rear panel will help to alleviate the situation. For example, if we mount the rectifiers and the regulator MOSFETs directly on the front panel, then the amplifier MOSFET junction temperature becomes

22°C + (282W x 0.26°C/W / 2) +10°C + 3°C + 27°C = 99 °C, only a slight improvement.

The average heat sink temperature is then 59°C with 22°C ambient.

The heat source distribution used in the experiment is, from the thermal point of view, far from optimum. By rotating it 90° about the centre of the heat sink, for example, the heat sources are wider apart laterally to make better use of the width of the sink. Furthermore, we can place the heat sources further away from the top than from the bottom edge, thus giving more dissipation area around the sources in the vertical direction. Both of these factors will lead to a lower peak temperature, even though the average temperature will remain the same, as the following simulation result shows. The maximum temperature difference over the heat sink is reduced from 20°C to 7.3°C just by clever but realistic placement of the heat sources.

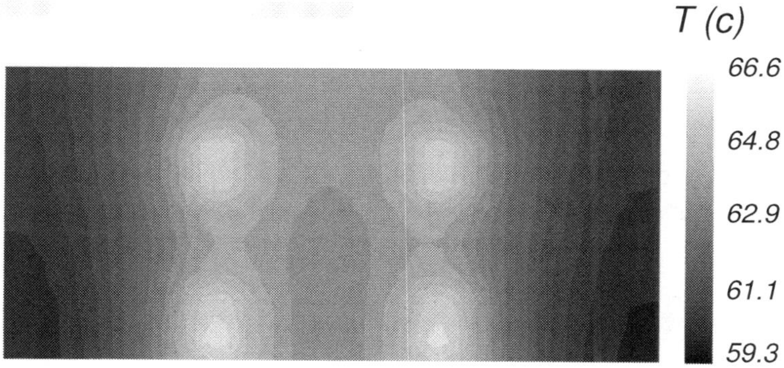

Figure 5: Reduced temperature differences with optimised heat source placement (simulated)

Just to confirm the accuracy of the simulation, the same configuration was measured experimentally after allowing the heat sink to reach steady state after 1 hour (**Figure 6**):

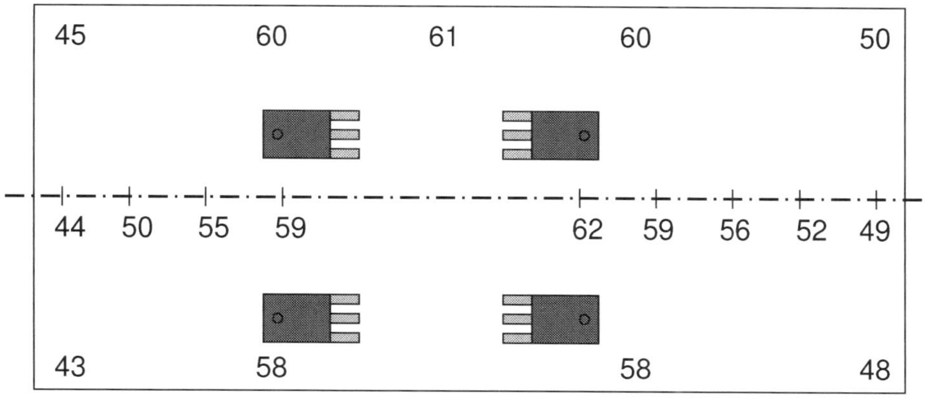

Figure 6: Reduced temperature differences with optimised heat source placement (measured)

The lower temperature measured at the free ends of the heat sink are contributed to extra heat sinking effect of the front and rear panels of the case.

You might be tempted to even out the temperature immediately around the heat source further with the aid of heat spreaders. However, bear in mind that such spreaders are only effective if they cover a large area, and have substantially less lateral thermal resistance than the heat sink base plate, which in this case is 8mm solid aluminium. Thus, an effective spreader has to be at least 12mm aluminium, or 6mm copper. And it is a misconception to think that bolting a heat spreader on is equal to adding thickness to the heat sink base plate monolithically. This is certainly not the case, due to the finite flatness of the interfacing surfaces, and hence the necessity to use a conformal interfacing medium, such as thermal wax. Such a medium has easily 10x higher thermal resistance than the parent metal itself. Only by soldering a heat spreader of the same material directly onto the heat sink you will realise the full benefit.

Fine points to note

So far so good. But you can easily lose more performance if you are not careful. For example, the calculated values here apply to heat sinks with vertical fins free from obstructions at the top and the bottom. Turning the heat sink 90° such that the fins now become horizontal will increase its Rth by 10~15%. Similar loss in performance is to be expected by placing the heatsink horizontal (exceptions are heat sinks that are designed for such orientations, such as finger coolers for CPUs), and by using clear anodising instead of black anodising. Another surface treatment sometimes found on heat sinks is powder coating. It is hard to believe that there is no performance penalty by using powder coating instead of anodising. A typical anodising layer is somewhat less than 10μm thick consisting of porous aluminium oxide with a thermal conductivity of about 16 W/m/K. A typical powder coating is somewhere between 50~125μm thick with a conductivity of 0.2 W/m/K. While its contribution to the resistance of the total thermal path might not be dominating, it is still a factor of 400 worse than anodising. One cannot help to think that the advocates of powder coating might have a hidden agenda as powder coating is much more effective to hide cosmetic defects in extruded aluminium [5].

Other thermal design issues

For optimal deployment of the heat sinks, it has already been mentioned that the fins should remain unobstructed. The heat sink back surface, as well as inner surfaces of the front and rear plates, also constitute additional heat radiating surfaces. These should therefore also receive minimum obstruction. The best way to do so is to leave the bottom plate clear from the heat sinks. There is no real reasons to fear large openings at the bottom unless you have small house pets (lizards, mice and cockroaches included), small kids who could craw under the 12mm gap between heat sink and floor, and the like. The top plate should optimally be just a wire mesh or perforated plate with >50% opening ratio.

But they would vibrate, you might say. They would, since they are thin and light. But because they are highly perforated, the vibration is likely to come from mechanical coupling through a heavy amplifier enclosure rather than acoustic pressure waves bouncing back and forth in your listening room. And because they are highly perforated, they are also not effective as vibrating membranes which will generate disturbances to your listening experience. In this respect, wire mesh is actually a better choice, as the internal friction between the large number of wire crossings is highly effective in damping any vibrations. In any case, magnetic materials should be avoided for the enclosure. They can easily be set into vibration when placed sufficiently close to the stray magnetic field of the power transformer(s).

There are many well known insulators one can use between the MOSFET and the sink – aluminium oxide, kapton, mica, even beryllium oxide which is now banned worldwide due to toxicity. The tricky bit is that they all come with different standard thicknesses, and many of them require transition layers on both sides, such as a thermal conductive wax. The proprietary Keratherm 86/82 [6] has also been mentioned above.

Table 1 below shows the temperature difference across different insulators, using the footprint of a 2SK1530 (21 x 26 mm), and 32W dissipation.

Material	Thickness (m)	Thermal Conductivity (W/m/K)	Transition layers (Aavid Ultrastick)	°C/W	dT @ 32W
Al_2O_3	0.001	25	2	0.13	4.26
BeO	0.001	330	2	0.07	2.10
Kapton	0.00005	0.24	2	0.44	14.13
Mica	0.00005	0.4	2	0.29	9.25
Keratherm 86/82	0.00025	6.5	0	0.07	2.25

Table 1: Thermal insulator temperature gradient for different materials

As one can see, the best performer is BeO, now unobtainium. This is closely followed by Keratherm, not because it has the best thermal conductivity, but because it is compliant and does not require transition layers (thermal wax such as Aavid Ultrastick [7]). Only then comes alumina, then mica, and Kapton is actually pretty useless despite its high price and glamour factor.

There are more and more such ceramic filled elastomer insulator sheets appearing on the market in the last few years. So Keratherm is by no means the only or the best. But the above illustrates clearly that the complete picture is more important than single material parameters.

Let's look at device clamping next. For Keratherm-like insulators a tensile strength of 10 N/mm^2 is specified. Even though compressive strength is normally higher, it is a good indicator for the maximum amount of pressure one can apply. For a 2SK1530, this converts to 5000N !! A single M3 bolt in high tensile steel can exert 3800N. But if you want non-magnetic stainless steel, then it reduces to 1750N. But does such a high pressure help? Bergquist shows [8] that depending on insulator types, there is performance to be gained by applying pressure as high as 1.3 N/mm^2 when using elastomeric insulators. It is less so if a thermal wax is used as a transition layer. E.g. only 0.6 N/mm^2 is recommended for Aavid Ultrastick. This is probably due to the fact that the thermal wax will change to liquid phase at some 70°C, and thus will then adapt itself to the exact shape of any gap between device and sink.
The problem with single-screw mounting is that the pressure is applied at the wrong place, i.e. not directly under the silicon substrate. It is normally (and understandably) at the top of the package above the chip itself. Thus, possibly less than half of the applied pressure is seen by the chip, and the pressure decreases rapidly with distance from the mounting hole. Therefore, even if you don't intend to apply the full 1750N, it is still beneficial to deploy an additional clamp directly over the silicon substrate, as shown in **photo 1**. The bottom surface of the clamp should have a slight curvature, or a thin elastomeric layer in between, to avoid excessive stress concentration which might in turn cause premature failure of the transistor encapsulation over lifetime.

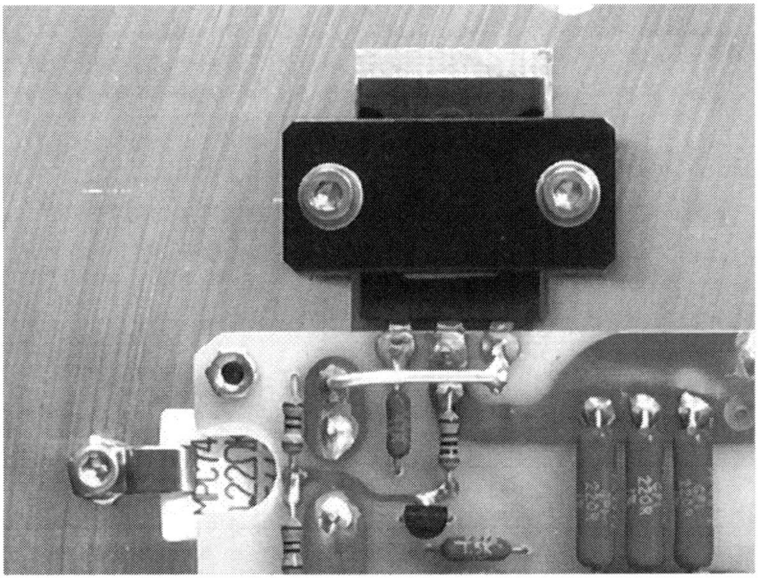

Photo 1: More uniform pressure distribution by use of a clamp.

Overall Dimensions and Enclosure Proportions

Beauty is in the eye of the beholder. And this applies equally well to audio gear. But for audio gear, the issue is complicated by the WAF (Wife Acceptance Factor).
Having said that, there is no escaping the fact that high power Class A amplifiers need large heat sinks, and people expect ´fat´ enclosures psychologically. Also high end gear always has thick front panels; thin front panels always look "cheap", irrespective of overall size.
So it is then a matter of proportions. An informal survey of commercial products suggests that the ratio of width to height should be between 2.2 to 2.7, and the front plate thickness should not be less than 4% of the width. The depth should be around 1x to 1.2x of the width. Too much depth means long heat sinks and it is difficult to spread the heat adequately.

Since the depth and the height are already given by the selected heat sink, the overall dimensions can be arrived at quickly. One only needs to make sure there is sufficient space for other components, notably the power transformer(s). And these should not come too close to the amplifier PCBs because of their stray magnetic fields. Transformers with a built-in magnetic shield help; Faraday cages in Mu-Metal are even better.
In the case of the F5X diyaudio enclosure design (**photos 2, 3, 4**), the inside dimensions are 250 W x 344 D x 140 H. This will take 2 F5X PCBs, 2x 500VA toroidal transformers (for dual mono), 16 power supply capacitors of 30mm diameter x 50mm height, plus enough space for wiring, slow start, protection circuits, and even voltage regulators with associated heat sinks.

Photo 2: FX5 enclosure design, top-front view

Photo 3: FX5 enclosure design, inside top view

Photo 4: FX5 enclosure design, one heatsink with mounted amplifier channel

However, one of the most impressive front panels the author has come across belongs to our Editor –it can only be described as a piece of art *(it is a piece of art! See [9] - Ed.)*.

Minimising vibrations

We already briefly described the use of perforated plates for top (and bottom) to minimize acoustically induced vibrations due to sound waves in the listening room. The other major source of vibrations is the power transformer. While over-dimensioning can reduce humming, the high pulse current

Photo 5: Vibrationally insulated power transformer sub chassis

experienced by all Class A power supplies still causes mechanical vibration at mains frequency. This vibration should also be decoupled from the enclosure, especially those made from thin iron plates. The best way to do so is to place the transformer(s) on a separate cradle with its own feet (**photo 5, 6**). The cradle is normally bolted to the amplifier enclosure during transportation. Once it is put in place, the bolts can be gradually loosened such that the cradle is lowered onto its own feet, leaving a physical gap between the cradle and the enclosure. *(This is similar to how turntable chassis are isolated from the environment in operation but bolted to the enclosure for transport – Ed.).* Both the cradle and the enclosure feet are lined with elastomeric material of some sort which provides vibration isolation and damping at 2x mains frequency. Vibration from the transformer now has to go through 2x such elastomers before it can reach the amplifier enclosure, the table or rack being an additional sink for such vibration energy.

Photo 6: Power transformer sub chassis in transport position

Editors' note:

The thermal simulations have been performed on-line at www.r-tools.com. *Sample thermal simulations are available at no charge.*

I am indebted to Zaher Aboumourad *of* Mersen Canada Toronto Inc *(creaters of R-tools) for kindly providing the thermal maps in this article and on the cover.*

References

1. http://www.diyaudio.com/forums/pass-labs/183362-f5x-euvl-approach.html
2. http://www.aavidthermalloy.com/technical/correct.shtml
3. http://www.semicon.toshiba.co.jp/eng/shared/reliability_pdf/bde0128g_chap02.pdf
4. http://www.semicon.toshiba.co.jp/eng/shared/reliability_pdf/bde0128g_chap03.pdf
5. http://www.alphaanodising.co.uk/technical%20info.html
6. http://www.kerafol.com/en/innovation-in-thermal-management/products-solutions/standard-films/keratherm-red.html
7. http://www.aavidthermalloy.com/products/options/ultrastick.pdf
8. http://www.bergquistcompany.com/pdfs/techLibrary/SilPad_Overview.pdf
9. http://www.linearaudio.nl/beenthere-1.htm ; also see http://www.helgijoensen.no/velcome.html

10 http://www.aavidthermalloy.com/technical/index.shtml

11. http://www.aavidthermalloy.com/technical/papers/pdfs/select.pdf
12. http://www.fischerelektronik.de/fischer/uploadfischerfcool/Fischer/A01_E.pdf
13. http://www.fischerelektronik.de/fileadmin/fischertemplates/download/Katalog/f.cool.e.pdf
14. http://www.r-theta.com/products_extrusion.asp
15. http://www.r-theta.com/rtools_front.html
16. http://www.r-tools.com/
17. http://www.federalpacific.com/literature/drytrans/10transformernoise.pdf
18. http://www.pmeweb.co.uk/audio/psu.pdf

A Headphone Buffer/Amplifier and Auto-EQ for Headphones

Richard N. Marsh

Before the iPOD and similar portable music products (MP3), even pre-Internet, we could listen with headphones through equipment, such as receivers and tape decks and there would be plenty of drive for any impedance headphone. The new portable products are very limited in output current capability and cannot drive many of the available headphones with low distortion. The iPOD-like products typically have a maximum output of little more than a 1 volt rms into high Z ear/head phones. A quick survey via the Internet shows that about ½ of all headphone models has an impedance of less than 100-200 Ohms. This is too low for most portable products to drive cleanly at adequate volume. Thus, an add-on buffer or high current driver for use with all headphones is needed.

Pre Internet era:

A stroll down memory lane – My history in DIY audio circuit development -

One of the most popular buffer circuits today is called the "diamond" buffer. The name comes from the bridge-like configuration of four transistors that when drawn looks like the playing card diamond. This circuit configuration was first described in the literature by R. Baker from MIT Lincoln Laboratory in a 1963 IEEE International Solid State paper (**Fig 3**). In 1975, I created a similar idea for my line level amplifier using a four transistor bridge circuit reconfigured for gain by swapping the emitters and collectors. Without knowing it then, my idea was somewhat similar to the arrangement like the earlier Sziklai compound transistor (1953**; Fig 1).** My idea was derived from the compound-complimentary circuit of **Fig.2**. This audio circuit was popular for many years. It was a natural take from tube designs and adapted for transistors. Making that circuit into a push-pull or diamond configuration was the idea I created for my own DIY preamp. It was later part of an article using a *passive* RIAA jFET phono stage and line driver I published in The Audio Amateur [5].

One of the advantages of the 'diamond' buffer configuration is that *if* all transistors have the *same* characteristics, the distortion is nulled or can-

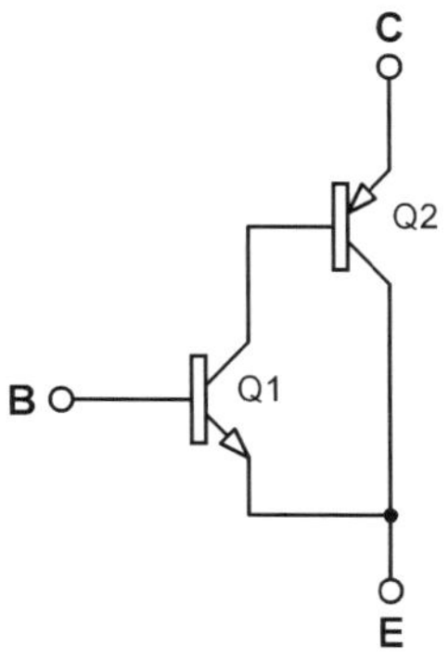

Figure 1: Sziklai compound Transistor (1953).

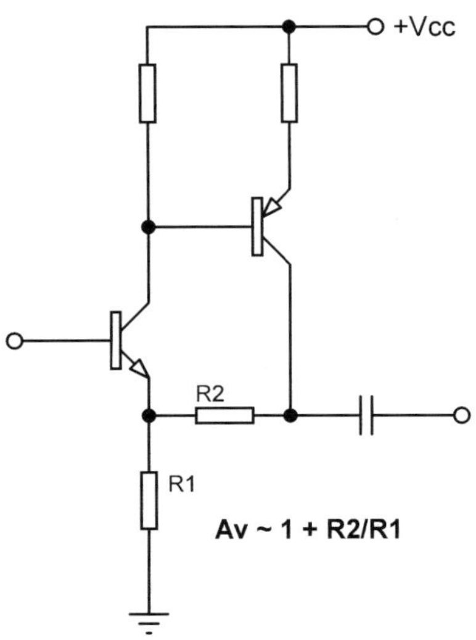

Figure 2: Compound-Complimentary with gain.

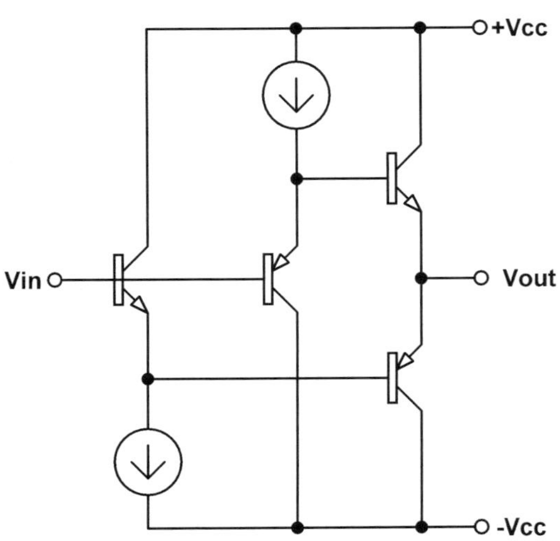

Figure 3:"Diamond" buffer (1963).

celled. After all, it is a bridge circuit topology. Configured as a line driver with a gain of ten, my minimalist 4 transistor topology with careful device matching was able to get below .005% THD but the output level had to be raised to 22 volts p-p output. At 1-2 volts rms output the THD was below my residual test limit of .0018% (calculated to be .0007% at 1 volt) using only 20db of negative feedback. You just have to use well matched compliments. I used a Tektronix transistor characteristic curve tracer for matching complimentary transistors. From that article, I got a call from several well known audio designers including an IC designer who asked about why I changed to this push-pull topology instead of the more conventional differential input – as I had used on the jFET phonograph stage. Primarily, I made that change because the line stage needed to handle larger signal levels with high linearity and because I only thought of it after I developed the phono stage. He asked if I had to do it over again, which circuit would I use in the phono stage. The bridge or diamond circuit, I said. Actually, I called it a compound-complementary push-pull circuit for lack of a better name. It could also be called a "diamond amplifier" circuit. The parallel input pair would be lower in noise potential than a differential pair circuit. I also preferred the arrangement because the input and feedback occurred in the same transistor – making the most direct correction. I discussed other benefits – low z circuit parts values I used meant the parasitic capacitances of the circuit layout and transistors would have very short time constants – this allows the circuit to have very wide bandwidth. My personal test equipment of that time could only measure a clean circuit rise time of 50ns. That was the same rise time as my generators' square wave rise time being used to drive the circuit! It was a very wideband circuit suitable for HF/RF. Later, I designed a cascoded and current sourced version to improve CMR and PSR. I didn't publish that circuit but offered it to anyone who contacted me. That design showed up later as a construction article in TAA [11] by one of those who had contacted me for the schematic.

With a circuit part substitution and implemented with a diamond buffer on input and output, a new circuit and patent was produced in hybrid form by Comlinear Corp in 1982 and the first IC version in 1987 by Comlinear and Elantec. It was called a current-mode feedback circuit and the rest is history. Many IC and discrete variations exist now. **Fig 4** is the original published topology which I used in my TAA [5] article. **Fig 5** is the basic changed new circuit for the Comlinear patent. In some ways, it is another derivative of that original basic "minimalist" 4 transistor diamond circuit with a high current linear output follower.

Let's get back to the new buffer being described here. Why a new buffer? I had hope that the IC buffer versions would be useful for ultra-linear audio applications. In 1979 I had tried the IC current buffer LH0002 in a line stage – a diamond or bridge configuration - and now tried a newer high current diamond buffer, the LME49600. They were very well matched for low *dc* offset but not as well matched for *ac* characteristics. The distortion from the LME49600 is very high when used *open*-loop. The printed specs for the very low THD being touted says it is used in the negative feedback loop of an Op-amp to get the distortion down. So, this device *must* be used inside the feedback loop of an (IC) Op-amp and that raises the cost and complexity of the project as really great audiophile IC Op-amps are expensive.

The buffer circuit here does not require high amounts of overall negative feedback to get THD below -100db (< .001%) Although I have nothing against high negative feedback per se, the extra complexity just isn't always needed. This simple circuitry has, as main features, a minimal number of transistors, low cost and high power output. All we need is a balanced/symmetrical circuit and matched compliments to get very low distortion. The work-around to matching transistors has always been to use more NFB as a more practical method for high volume

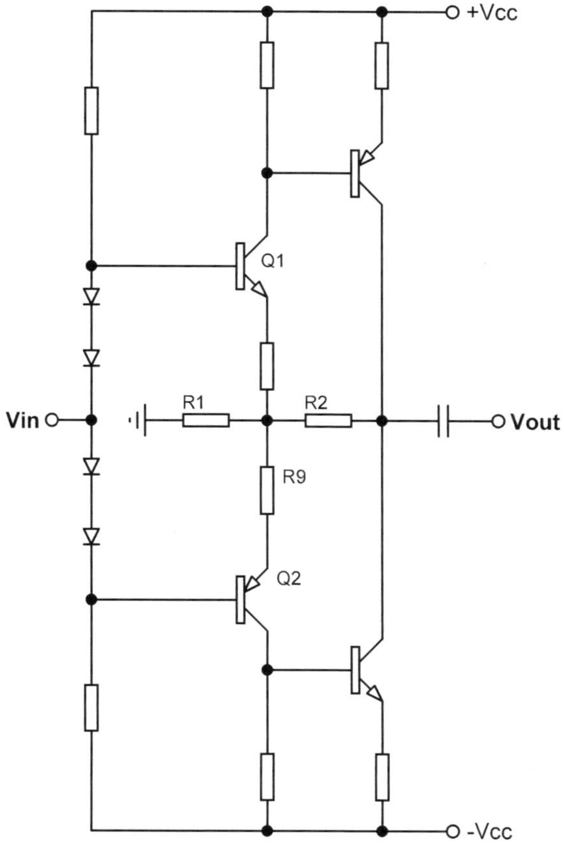

Figure 4: My C-C Push-Pull (1980) aka "Diamond" amplifier.

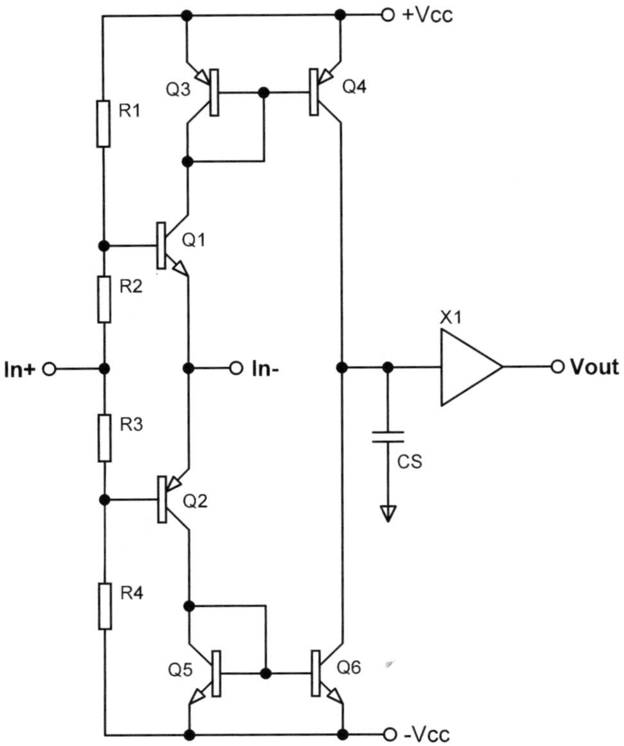

Figure 5: Comlinear current feedback (1982).

production.

If you remove the gain setting resistor (R1) from Fig 4 and reduce R2 to zero ohms you have the new buffer. As this turns out, during research for this headphone buffer/driver design, it is also a complimentary push-pull version of what is known as a Sziklai transistor pair. This pair of NPN and PNP was first described in 1953 and patented by G.C.Sziklai and is sometimes known as a "compound transistor" or "compound Darlington." Current gain is similar to that of a Darlington pair, which is the product of the gains of the two transistors. Some advantages are: The combination is more immune to thermal runaway. And, a Sziklai derived complimentary push-pull arrangement would be more immune to NPN and PNP mismatches. As an interesting aside, the Sziklai pair most resembles the IGBT equivalent circuit.

Fig 6 shows the new jFET/BJT complementary push-pull buffer. By using a jFET instead of BJT at the input stage, we can have a stable, low distortion source of bias voltage for the high current output stage. At this point it would be easy to add two complimentary transistors as followers to drive low z loads as well. However, with such follower circuits, as the current draw increases so does the distortion. It didn't measure much but we can improve on the situation a little more. We can use another Sziklai transistor pair as output followers/drivers to reduce distortion at low load Z and we are done (**Fig 7**).

In all the intervening years, I haven't found any reason to change topology so when I needed a headphone amplifier, I started where I had left off – the diamond or bridge or complementary push-pull circuit topology. Push-pull complementary topologies in a wide range of variations have become well established. The gain shown in **Fig 8** is set at 12db for iPOD and MP3 players. Using it from the direct output of a SACD-CD/DVD player or from a computer audio line out seems Ok at this gain level, too. Notice no servo on the direct-coupled output. You might wonder how I get away with direct-coupling from input to output. Years ago, when I was explaining to Walt Jung [4] about capac-

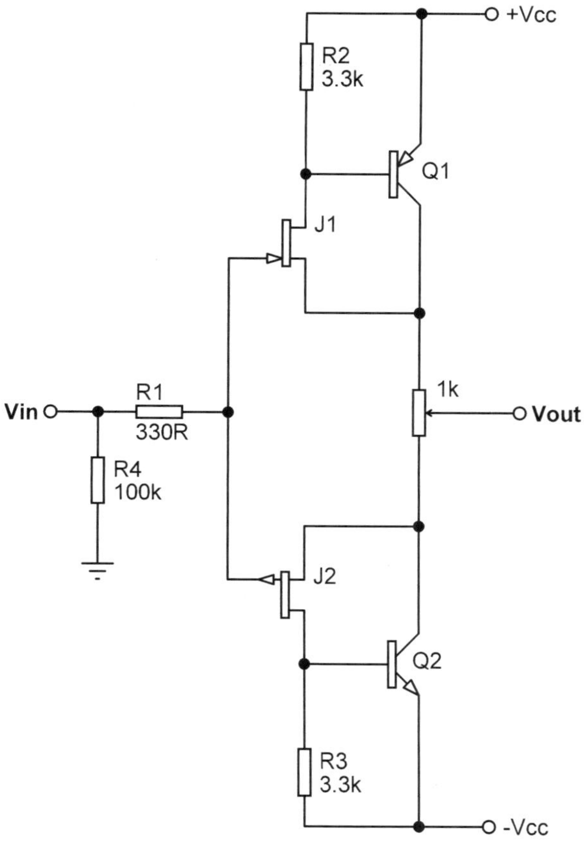

Figure 6: jFET-BJT complimentary push-pull buffer, suitable for driving into high Z loads such as an amplifier input, with extremely low THD.

itor design and I found that audible distortion was due to dielectric absorption, I told him that the best capacitor was no capacitor at all. But, he asked, what about dc offset and drift in the output? I said use a *servo* circuit to keep the dc nulled at the output. He applied my idea/concept of a servo to his own preamp design and that helped establish the "dc servo" as a standard feature when eliminating coupling capacitors in audio circuits.

This circuit, however, is very dc stable with time and temperature. Most of the discrete complimentary circuits and their derivatives in use today have multiple constant-current sources for biasing – one for each complimentary device. Consequently, these current sources and the complimentary transistors do not temperature track with one another very well. So, we still need a dc servo to keep dc from the output. I have had this prototype powered 24/7 for a month and I never see more than a couple millivolts drift away from zero; not even when I touch the transistors or from room temp changes etc. It helps to run the transistors warmer than room temp to minimize ambient temperature variations on the transistors. And, no more dc offset than 20mV from cold start to temperature stable operation (a few minutes) has been observed.

We can calculate the closed-loop Harmonic Distortion as follows. We have 62-12 = 52dB of feedback. Add -52 dB to the measured open-loop HD and we get a closed-loop HD of -52 plus -55 = -107dB for second harmonic and -117dB for the 3rd harmonic. Lighter loads (>30 Ohms) will give lower THD as would closer matching. The THD does not increase with increased source resistance.

Some might think that circuits with exceptionally low THD (more than -100 dB) are over-kill and not needed. After all, it is said, we can't hear it. I did some experiments years ago using electrostatic headphones and electrostatic loudspeakers which had very low distortion to determine what level of THD I could hear. I found with midrange frequencies at 90dB SPL I could not hear harmonics below 0.1%

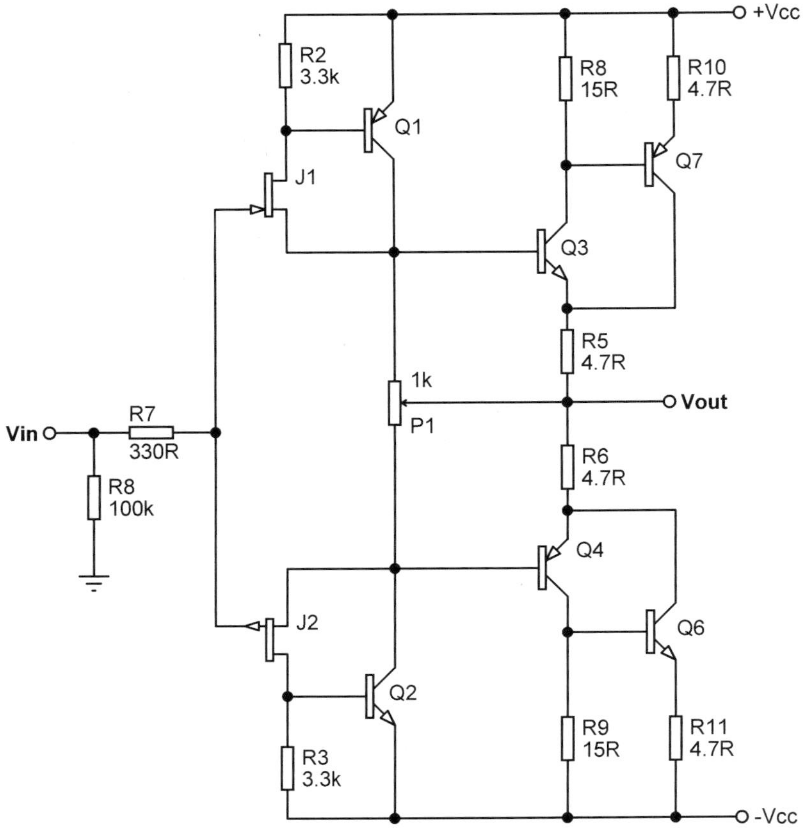

Figure 7: The completed Buffer

Open-loop gain	+62dB
Open-loop bandwidth (-3dB)	25kHz
Open-loop Harmonic Distortion at 1.0v rms, 1 kHz	
2nd harmonic	-55dB
3rd harmonic	-65dB
Open-loop noise (30 KHz BW and no weighting)	-85dbv
[Closed-loop gain]	+12dB
[Closed-loop s/n with input shorted]	-135dB

Table 1: Prototype amp data using a worst-case 30 Ohm load.

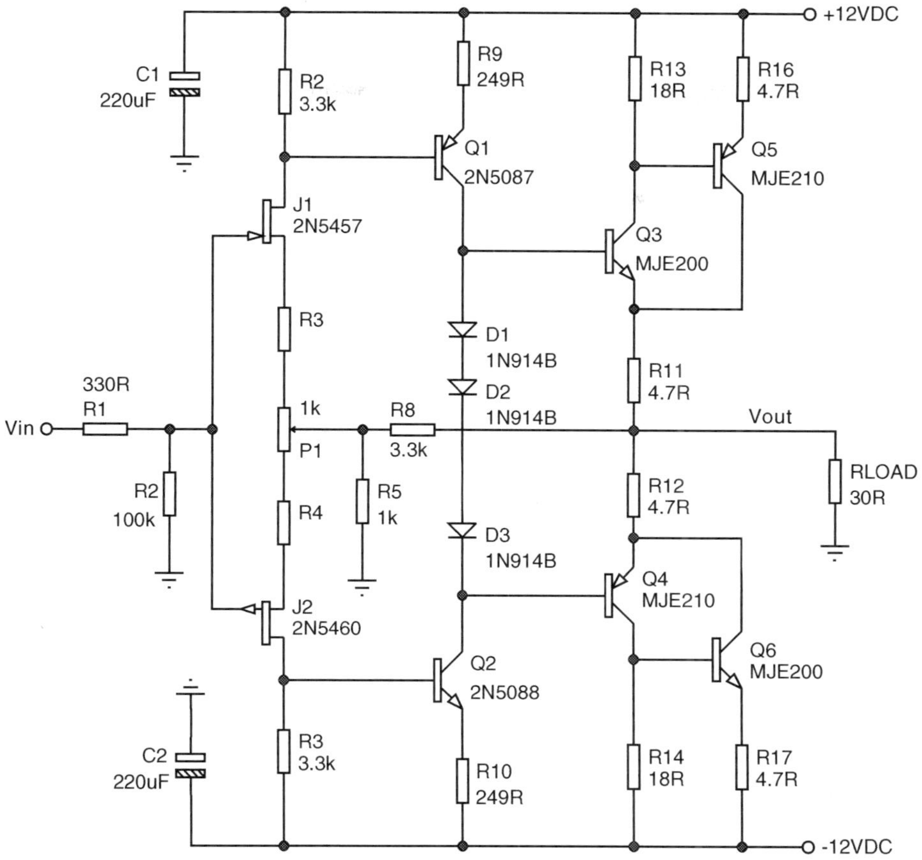

Figure 8: The complete headphone amplifier/buffer.

THD. So, I thought an amplifier with ten times lower THD or 0.01% should suffice for any circuit. However, that would only apply to a single amplifier and a single frequency. In the process of making a recording, there are dozens of amplifying stages the signals pass thru; maybe even hundreds of circuits. Consequently, I never hear music as accurate sounding as I have on playback of my own master recordings using just a flat response condenser mic pair, mic preamp and recorder. Not even close! I would put forward to you that when a full orchestra plays, there are hundreds (maybe thousands?) of tones that if each tone went thru just one amplifier with .01% THD, the total of all those added harmonics from each of those frequencies would amount to an equivalent of more than 1% THD. It is audible. This might partially explain the difference between commercial recordings/playback and hearing a direct master recording. So, the fewer the stages in recording and playback, the better it sounds but it also means we really need to get amplifier THD down even more to stay under the accumulated audible effects of THD and that would be at least .001% or better.

The Headphones

What good is a low distortion signal when you still have to listen to 'lumpy' frequency response errors from the headphones? Very few headphones are flat in frequency response at the ear canal. Can anything be done about that? Fortunately, the room correction products, such as those made by Audyssey are now included within low cost AV receivers and they can be used to automatically adjust the headphone response to flat. Or, you could use a separate stand alone auto-room EQ system. I will tell you how, later.

In the past we had to use graphic equalizers with microphones and swept sine waves or spot frequencies and plot the response on paper. Or, measured with filtered octave and1/3-octave bands of noise. There was no consumer RTA available when I started doing room measurements and speaker measurements. So, I conceived of one and with my coworker, Bob Jones, we designed and made our own RTA that connected to an oscilloscope for the display and we were ready to go. That design was the very first consumer RTA and was published in Popular Electronics, September and October 1977, as a DIY project for the advanced constructor. Soon, afterwards there became many consumer products with RTA displays (using LED bar graph displays) which still exist today. But, now we can download an application for an RTA into our iTouch or iPad and have the same thing for less work and effort and low cost. And, much smaller than a rack mount chassis full of parts.
However, here is how I do it now — There are a couple ways (**Photo 2, 3**). Both ways worked well. The first way is to take a small diameter microphone – such as the ¼ inch diameter Panasonic electret cap-

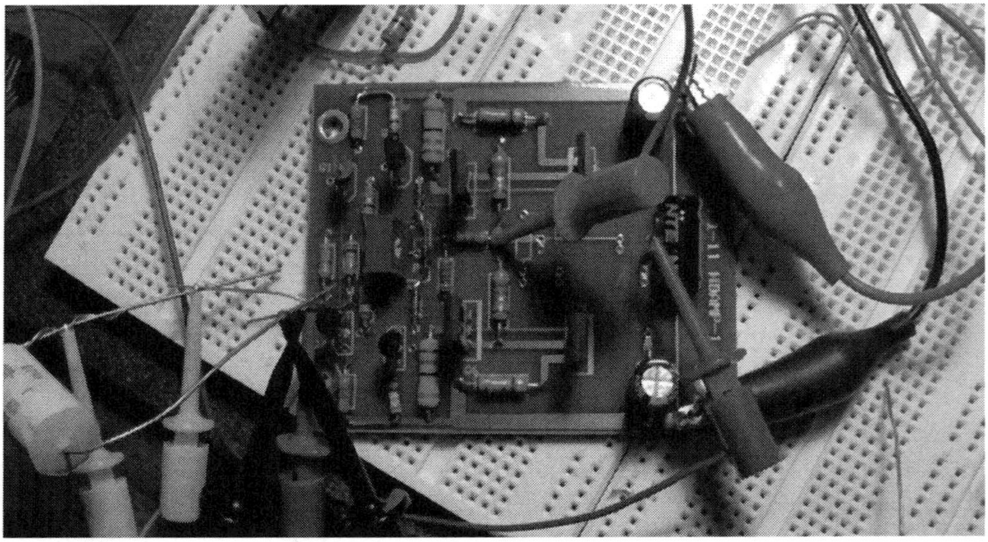

Photo 1: the PCB for the prototype was made by ExpressPCB.com; signal and power ground planes are on the bottom side.

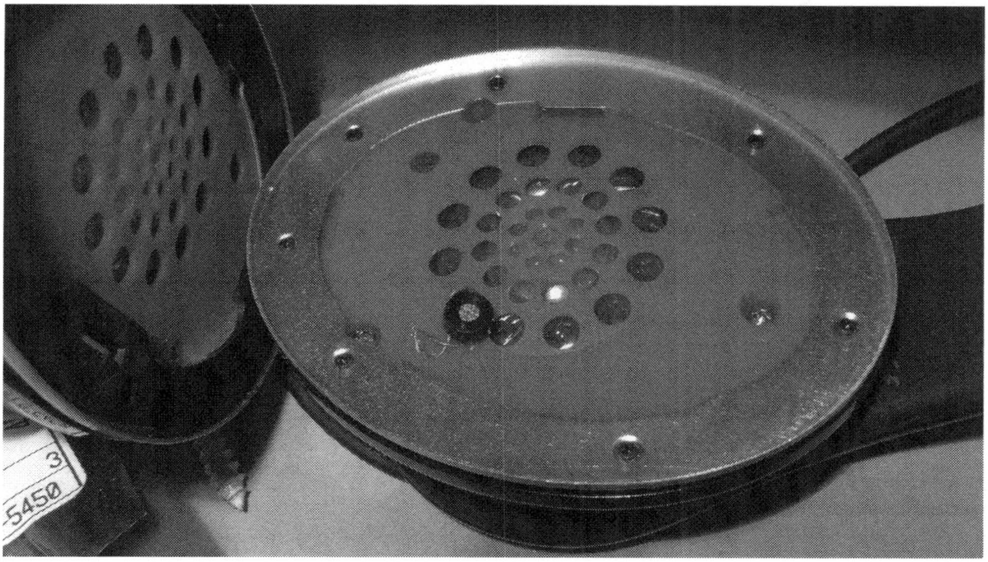

Photo 2: Use a small electret microphone capsule attached inside the headphone's ear cup.

sule and mount it inside the headphone ear cup and with fine gauge wires attached to the capsule; they come out to a cable and plugs into the AV receiver's microphone Auto-EQ input. Wear the headphone and run the auto-EQ response correction as per the receiver's instructions and you will have

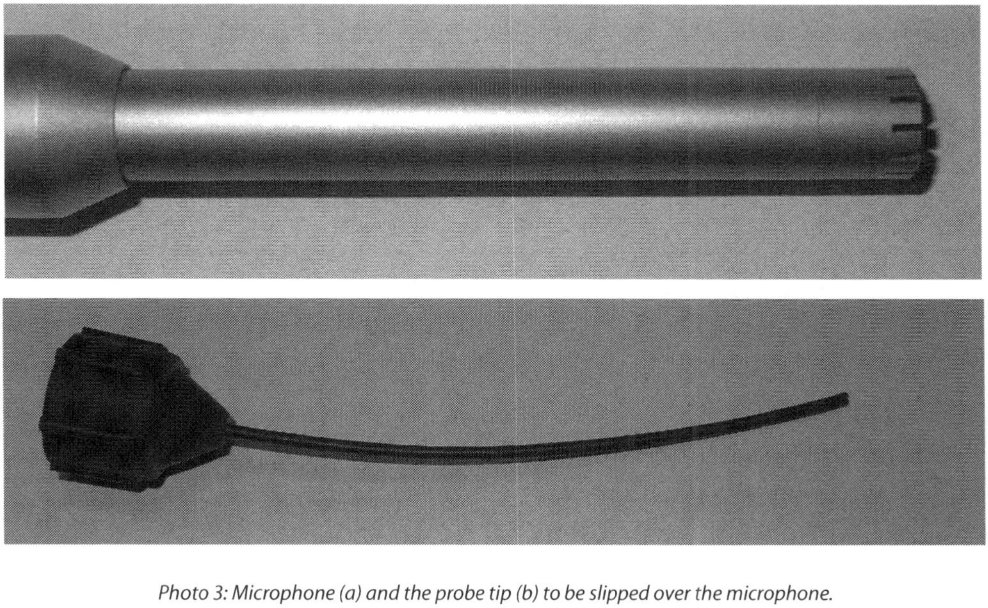

Photo 3: Microphone (a) and the probe tip (b) to be slipped over the microphone.

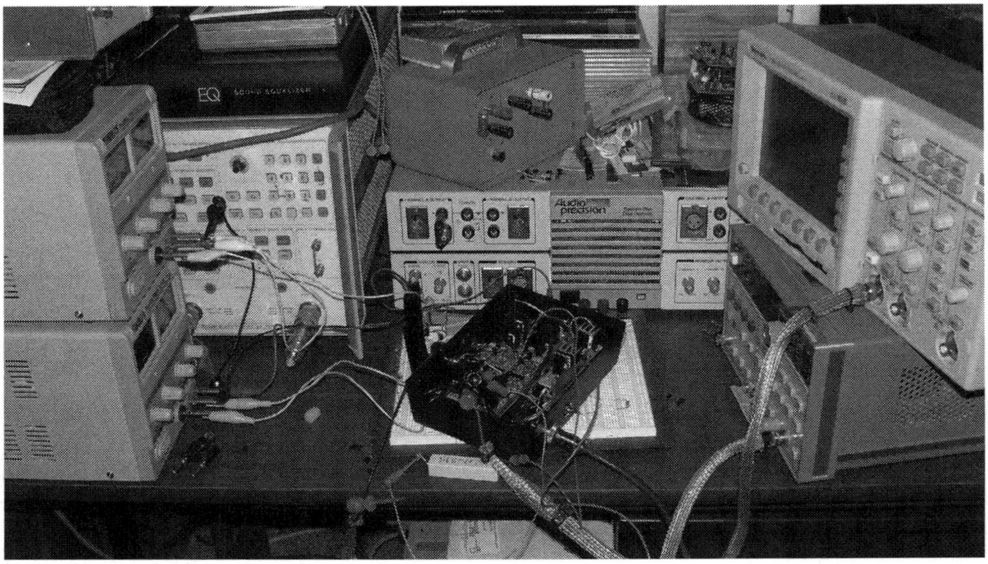

Photo 4: The prototype headphone amplifier being tested.

a pretty flat response using your own ears and your headphone of choice. If you have a separate or 'stand alone' auto-EQ system (Audyssey, XTZ, ARC etcetera), you can use the microphone included in that system or one of the many high quality microphones advertised and fit it with a 'probe' tip. Make a 'probe' adapter as shown in Photo 3 [14] and place it on the end of the microphone (a tube of .05 inch inner diameter) and place it between your ear and the headphone ear pad and run the auto-EQ system. Any minor corrections in microphones for flatness can be done but the microphones are so flat compared to any headphone that it will be a big improvement anyway. These approaches flatten the whole path – ear shape and size, headphone, air seal/leakage and cable/amp responses. It is fast - so a new auto-EQ can be quickly done with someone else using the headphone.

Make your microphone "probe" fit over a 1/4 or 1/2 inch microphone end, tightly. The probe tip then goes under the headphone ear pad (while you are wearing the headphone) to receive the acoustic signal and send the response to the Auto-EQ system. The data shown next (**Fig.9**) is a before and

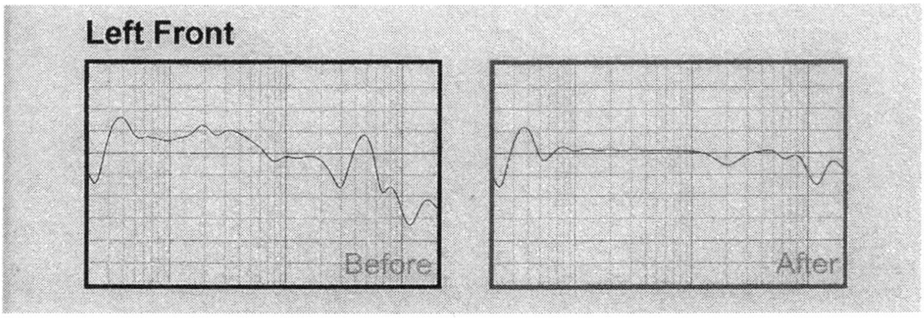

Figure 9: Before and after auto-EQ of the headphone response.

after result of a single pass with an Auto-EQ system made by Audyssey. In this case, a well known brand of headphone in the $300 dollar class is shown being corrected via a probe microphone using a PRO Audyssey MultEQ-XT system.

No attempt to make it better (flatter) was tried - which could be done - it's just to show you the concept and that it does what is expected. The probe tip adapter could have its response compared to the non-probe microphone and any differences noted and used in correction of the final response.

I had the opportunity to compare the headphone results against an expensive B&K dummy head system and using a B&K 4182 probe microphone and the agreement was very close. I believe the more accurate is the one shown here as it used real ears (mine) and thus can be individualized for each user.

So, here we are with a great little amplifier for all headphones and a method to obtain better frequency response from your headphone, as well. The combination provides outstanding results at low cost. With higher bias and adequate heat-sinking, this circuit has been operated up to about 12 Watts and will then drive 8 ohm loads at low distortion.

I hope that this headphone EQ approach will become an integrated package that is available everywhere in the future.

Enjoy !!

References and Further Reading

[1] Knowles Technical Bulletin TB15 "Probe microphone kit application notes."

[2] Marsh, R.N. – "How Linear distortion can sound non-linear". The Absolute Sound (TAS) #48, 1987 pg105-107, The Cutting Edge.

[3] Marsh, R.N. – "Direct Coupled Designs." TAS #55, 1988 pg. 97-99, The Cutting Edge.

[4] Marsh, R.N., Jung, W. - "Picking Capacitors", Part I and II, AUDIO, Feb & March 1980.

[5] Marsh, R.N. – "A Passively Equalized Phono Preamplifier." The Audio Amateur (TAA); issue 3/80, June 1980.

[6] B&K probe microphone Model 4182.

[7] Marsh, R.N. – letter on Servo Control for Elimination of Amplifier Coupling Capacitors, TAA, Issue 5, 1984.

[8] Marsh, R. N. – "A Complementary Push-Pull Power Supply" TAA, Issue 1, 1988

[9] Marsh R.N. – "Power Up: An Overview of Power Supply Considerations" TAA, Issue 3, 1983

[10] Marsh, R.N. – "POOGE 3" TAA, Issue 4, 1985

[11] Marsh, R.N. – "Understanding Common-mode Signals" AUDIO, Feb 1988

[12] Marsh, R.N., Jones, B.A. - "A Real Time Audio Analyzer", Popular Electronics, Sept 1977.

[13] Marsh, R.N. – "Resonances: Getting Beyond Frequency Response". TAS, Sept/Oct 1990

[14] Linda P. Franzoni, Christopher M. Elliott - An innovative design of a probe-tube attachment for a 1/2-in. microphone, J. Acoust. Soc. Am. **104** (5), November 1998.

Power supply

The power supply is not shown as it can be any good, fast, quiet, stable, regulated circuit that provides 12-15vdc at >200mA per rail. There have been many great circuits published elsewhere – including the Internet. To speed up the fab time, the prototype power supply used an excellent Plug-N-Play regulator circuit of series type premade by Belleson – SPJ series. However, I prefer the shunt regulators or a 'splitter' type circuit. These are also being called virtual ground circuits. They inherently keep both + and – voltages the same.

Rail splitting wasn't new then but using high performance shunt regulators in audio was new. I showed a simple example of a shunt regulator/rail splitter in TAA [8]. The subsequent TI part that came out afterwards does the same thing but the maximum current output is too low for these circuits.

While I am on the subject of power supplies, let me bring up another simple yet very effective way to keep the + and – rail voltages the same; Namely, by placing a large value electrolytic capacitor (hundreds to thousands of microfarads) across the rails so the supply voltage will be made the same on each rail under dynamic load conditions. The voltage may vary in magnitude but the two voltages will be kept equal. This is especially useful and practical for power amplifiers where a high current, regulated voltage cost would approach that of the amplifier itself. It is more important to amplifier performance if you keep the dynamic voltage differential of the supply rails to a minimum than what their absolute values are. It also reduces the burden on the amplifier to reject voltage changes on the power supply lines (PSRR). I have used this for twenty years and commercially with great success.

I have not shown film bypass capacitors but, of course, they are always used for highest performance. As I found and first reported decades ago, the primary culprit is the dielectric film used in capacitors which has the largest effect on sound .... Polystyrene, Teflon and polypropylene are the best for audio. They have the lowest dielectric absorption characteristics which I found to cause audible degradation using <u>*asymmetrical*</u> waveform tests. Symmetrical sine wave tests tend to average out to zero or very low levels. But asymmetrical waveform tests showed 1-10% peak distortion with polar capacitors (mostly second harmonic); see for instance [2] and [4]. At least a minimum of 0.1uF should be used across supplies and electrolytic caps. Often even greater values are best. And, they should be placed at the amplifier circuit and not at the power supply if they are to do the most good.

Device matching

All BJT transistors should be selected to match their compliment in Hfe and Vbe to get the very lowest THD and dc offset voltage and minimal drift. But, I did a lot of testing of various transistors so you won't have to worry too much – even without tight matching, using these transistor part numbers, the harmonics will generally be more than -90db down from fundamentals with a load of 30 ohms. It is primarily second harmonic. No harmonics above the third were seen. The cost of any transistor used here was in the 10-15 cent range for small quantities via the Internet.

The best way to do complimentary matching is with a characteristic curve tracer (e.g. Tek 576). The second best and the easiest way to do the matching of Vbe and Hfe with the bipolar types is just to buy a small hand held portable tester from PEAK Electronic Design Ltd. – Model: Atlas DCA55. I found it on eBay for 50 dollars or less (new). The headphone amplifier THD result shown here (> -100db) is from using this simple tester for device matching.

The 1K pot is for DC trim on the output.... If you have matched your compliments well, the trim will be a minor adjustment to zero on the output. Or, you can replace it with fixed resistors. And, of course, the four output transistors (Q3,4,6,7) need a heatsink for their TO220 package.

The only issue to really deal with are the input jFET's. These devices are not like the devices 2SJ74/2SK170 pair. Those are large geometry devices – needed for low noise - which means high capacitance; highly *non-linear* capacitance. Usually an additional cascoding transistor is required to deal with this capacitance and the distortion created in interfacing. The jFET's used here have very low capacitance. Also, they do not have as wild parameter variations from piece to piece. But, the spread in Idss is still something to deal with if you want the lowest THD and output dc offset. Fortunately, you won't have to pay a lot to get them and finding a matched compliment doesn't take a truck load of parts. A simple test jig will do the trick for you (**Fig 10**). You want two complimentary jFET's that measure close to .8 mA each or 0.8 volt measured across a 1K source resistor. This will put the devices close to their zero TC operating point. You decide the tolerance you want but 5% or better is good and fairly easy to obtain. Expect to get about 20% matches from a batch which will meet the 0.8 volt requirement. I try for better matching which takes more time and samples because I want the absolute lowest possible distortion and dc offset and drift. I also use a Tektronix curve tracer to get exact matches over a range of currents and voltages.

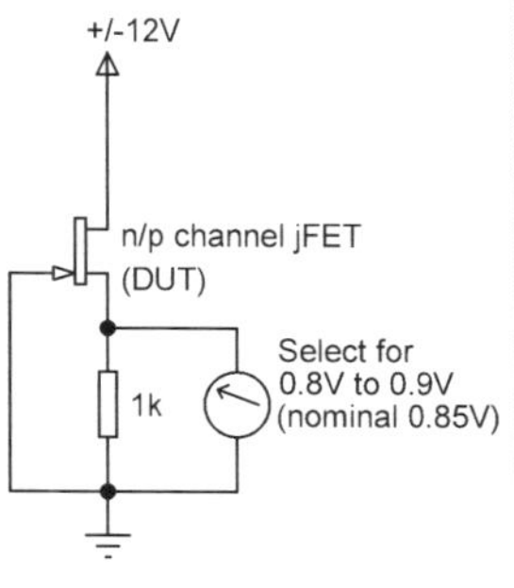

Figure 10: jFET matching jig.

DIY Low-noise Microphone Preamplifiers, Part II

Scott Wurcer

A few general comments

I have not been religious about showing all supply bypassing in the following schematics; the builder should use good practice in layout and supply bypassing. I leave the issue of component quality to individual choice, but note that I have made an effort to eliminate high value capacitors in the signal path as a nod to one of the hot buttons of the audiophile community. Care should also be taken to observe correct output phase whenever possible, positive air pressure should cause a positive polarity output.

Also, note that the present article is in many ways a continuation of my article on the same subject in Volume 1 so you may want to review that [1].

Microphone Capsules

Microphone capsules are a subject that I could not do full justice to for several reasons. Since they are a mechanical transducer there are many non-electrical issues involved in their design and performance. The offshoot of this is that they often have a distinctive "voice" which invariably appeals to some and not others. The noted recording engineer Harvey Gerst has stated in an interview that "perfect" instrument grade microphones often make lifeless recordings. Early important recording microphones also frequently used vacuum tubes and/or transformers, both of which can contribute their own "color" to the sound and be very costly at the highest quality level. I don't want to address potentially very subjective issues, and the reader is encouraged to learn and experiment on their own. There is a wealth of information available on choosing microphones, equalization, mixing, recording techniques, etc. For the same reasons I will not discuss here the recreation of these classic microphones.

There are three basic families of capsules readily available to the DIY community. There is a large number of fully integrated capsules (JFET included) and I could not possibly list them all. Panasonic's WM-61A and similar are popular small capsules that many DIY microphones are designed around. They usually have a very low I_{dss} JFET (200-600μA) so they work well with plug in power (PIP) which is found in computer headsets/microphones, low cost recorders, and other low end applications. The low cost and wide availability of these capsules makes them very popular with DIY'ers, but their relatively poor noise and dynamic range are limitations.

These capsules can often be modified by cutting the connection to ground on the back of the capsule [1], which enables the JFET to be configured as a source follower lowering gain but increasing dynamic range. One thing that might not be immediately obvious is that the capsule capacitance can now be used as a coupling capacitor allowing one to inject a signal which might be useful in testing a circuit for dynamic range or distortion. This also enables one of these types of capsules to be used in the last circuit I will present where feedback is applied to this terminal. There is one large drawback though: the case is usually designed to be connected directly to ground so there is an opportunity to pick up hum and noise if it is not, making a double shield necessary.

There are some slightly higher end versions of these capsules, for instance, the Primo EM-172/3 which offers better noise performance. As an experiment (not a practical idea for a microphone) I took two EM-172's and connected them face to face and subtracted their outputs thinking that if I could get everything perfectly balanced their uncorrelated noises would add (root sum squared i.e. 3dB) but the ambient noise, common to both, would be rejected. I am including the result because it is a dramatic demonstration that the noise of these small capsules is a property of the external acoustic "circuit" and not the electronics. The plot, **Figure 1**, shows the differential output in balance with a barrier (a piece of electrical tape) between the capsules and then with the capsules touching, in both cases they are tightly bound together around their circumference. Electrically there is no difference in either case so in both the JFET's and resistor's noise contribution will be the same (+3dB of one side). The difference is dramatic, almost 10dB in places, which means the JFET contributes less than 10% of the noise (probably less). It is not easy to compute the exact acoustical circuit but the lower of the two must be a maximum for the electrical noise contribution.

The next type of capsule is the electret without internal JFET which is mostly sold into the OEM market. They are often available separately as replacements or to offer multi-pattern options, but there are several notable ones available in small quantities. Transound capsules are available in small quantities from JLI Electronics [2]. Notably the TSB165A but some of their capsules are not listed on their web site and require directordering; this includes the large diaphragm TSB-2222A.

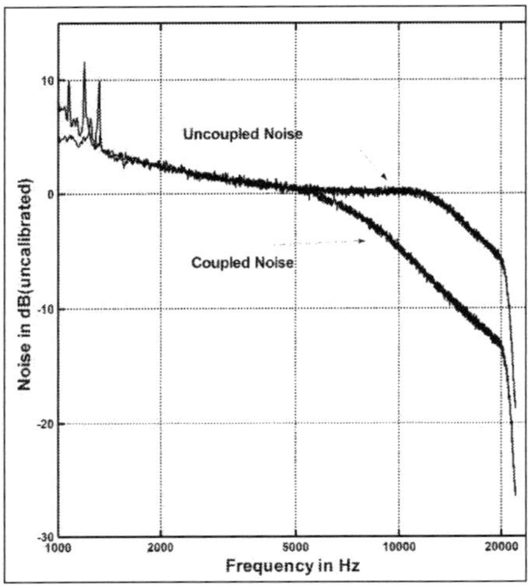

Figure 1: Demonstration of the Acoustic Nature of Capsule Noise

Selected Primo capsules are available from JLI and at least one hobby outlet, but unfortunately the machined bodies of the EM-21/3 make them special to OEM's and they are fairly expensive, but they are still listed as sometimes available in small quantities at ~$60 [3]. People frequently purchase used or discounted microphones to get a capsule and the mechanical components with the goal of replacing some or all of the electronics. I obtained mine from used Nakamichi CM-300 microphones from which I also got a body and integral XLR connector.

The third major class is the classic large (~1") externally polarized microphone capsule. There is so much lore surrounding the history of these and there is so much discussion of them both in print and on the web that they deserve a separate article. I only have the opportunity here to show some options for incorporating them in the circuits I will present; functionally the only difference is that these capsules require an external bias voltage for operation. The reader is again referred to the excellent overviews offered by the major microphone manufacturers for more information.

There are several sources for these capsules available to the DIY community, the most economical being microphone-parts.com. They offer the two major versions (center terminated and edge terminated) with dual diaphragms which enable programmable patterns. Their site contains a wealth of information on the differences and applications of these capsules. In a later section I will give several options for generating external capsule bias with little or no increase in power consumption.

Selecting JFET's

The JFET is the preferred impedance converting element in most solid state microphones. In most cases one would want to verify in advance that a JFET has low noise and V_p and/or I_{dss} in the proper range. **Figure 2** is a very simple test circuit for this. The battery floats so positive drain bias and negative gate bias are both created from it. R_2 is selected to give 5V drop at the desired test current and R1 is adjusted until the output reads -5V. This should produce enough gain to overcome the input noise of the measurement system. I used a good external sound card and Adobe Audition to plot the noise spectrum. To calibrate the measurement simply switch a 604kΩ resistor in series with the gate, this provides a 100nV/√Hz reference level. Don't worry if the resistor noise is rolled off at the high end due to the Miller multiplied C_{gd} of the JFET; the resistor noise is flat with frequency and we just need one point as reference level. 100nV will swamp the high frequency noise of any JFET, so simply normalize the plot of the JFET noise to make the 100nV level 40dB and the noise will read in dBnV or 0dB is now 1nV. If your plotting software has the capability you can simply make the vertical axis read directly in nV. Simply flip the battery to measure P-channel JFET's in the same way (the DC output is now positive). This calibration technique also eliminates the need to factor out the equivalent noise bandwidth, any program that can display the noise spectrum in usable with the same calibration.

The same fixture can be used to extract the V_p ,I_{dss}, and β, the transconductance parameter. Measure the V_{gs} at two well-spaced values of I_d say 5:1 and use the basic JFET equation to com-

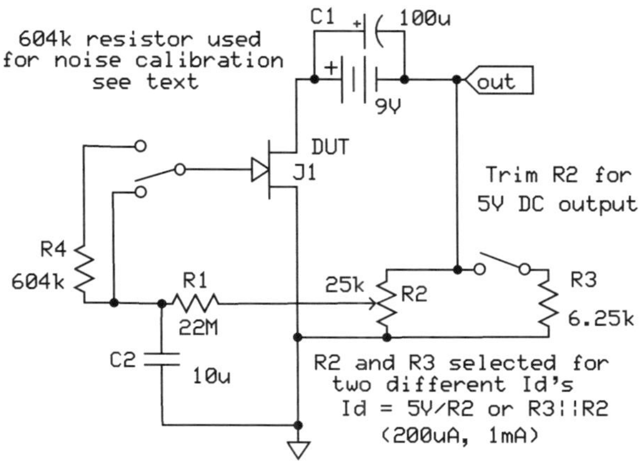

Figure 2: JFET Test Fixture

pute the values. If you want to be more precise, change R2 so that at each current the V_{ds} will be approximately the same. The process is as follows:

1. Measure V_{gs} at two values of I_o (for example 20oµA and 1mA)

and since, $I_d = I_{dss}(1 - \frac{V_{gs}}{V_p})^2$

the ratio gives, $\sqrt{\frac{I_{d1}}{I_{d2}}} = \frac{V_p - V_{gs1}}{V_p - V_{gs2}}$

2. Solve for V_p then substitute back and solve for I_{dss}

and then $\beta, \left(\frac{I_{dss}}{V_p^2}\right)$

The transconductance is given by $gm = 2\sqrt{\beta I_d}$

These values will probably be accurate enough to pre-compute values for the bias resistors in most of the circuits here, or of course the suggested trims can be used.
It would be difficult to make an exhaustive list of JFETs useful in these circuits, I will try to list enough here so that anyone should be able to find a suitable choice. It also bears repeating that the input JFET will rarely (if ever) be the main source of noise in anything but possibly large capsule microphones.

N-Channel JFETs

2SK30, 2SK118 , 2SK117– Low I_{dss} often used in older microphone designs where they are biased at I_{dss}. Appears in many large diaphragm Neumann "clone" mikes and the classic Nakamichi 300 series.
J200 series – Can be selected for low I_{dss} and should substitute for the above.
2SK222 – Low I_{dss} and high gm (low noise) best tradeoff but hard to find.
J305, J309, J310, 2N4416 – switches and high gm RF FETs, low capacitances and good gm, may require noise or V_p selection for operation at low voltages. J310's from eBay @ 10 cents yielded very well to low noise.
BF862 – RF FET very low $V_p$and high I_{dss} with excellent yield to low noise, SMT only. It may need a ferrite bead in the gate lead to suppress parasitic oscillation (f_{cutoff} of 750MHz).
2SK170 – Classic low noise JFET but high parasitic capacitances, use with caution.

P-Channel JFETs

2SJ74, 2SJ107, LSJ74 – High gm and low noise but very high parasitic capacitances, not recommended in the input stage.
J175 – switch, possibly poor yield to low noise.

Battery vs. Phantom Power

The microphone circuits in this article were for the most part designed around 100 hours operation on a single fresh 9V battery. This would mean an uninterrupted recording period that would cover most applications by a large margin as well as compatibility with virtually any recorder. The same circuit could also run on an external supply, an easy solution could be a small USB "wart" using an LM2681 voltage doubler to make a +10V supply.

Still it may be undesirable for one reason or another to use batteries in the microphone, most importantly battery replacement during a critical recording session is obviously unacceptable. For this reason professional recording equipment uses a phantom power arrangement where the DC supply and signal share the same path. Standard phantom power, **Figure 3**, mates very well with two-wire + shield balanced signal transmission which is also ubiquitous throughout the professional audio industry. On the left in figure 3 is the fully differential connection and on the right is the pseudo-differential connection. The 48V standard for microphones was first proposed in 1966 by Neumann and was based on the wide availability of 48V emergency battery supplies for lighting and telephone systems. The 48V is typically supplied through two 6.8kΩ resistors (6.81kΩ is the closest standard 1% value). Many portable recorders now provide phantom power but in general battery life is greatly reduced when this feature is used (all current to the microphone is supplied at 48V) so a microphone that operates on a separate 9V battery again becomes an attractive option.

The DC currents must be equal in each leg because in some applications center-tapped transformer coupling is used and there must be zero net DC flux to insure low distortion. We have at the micro-

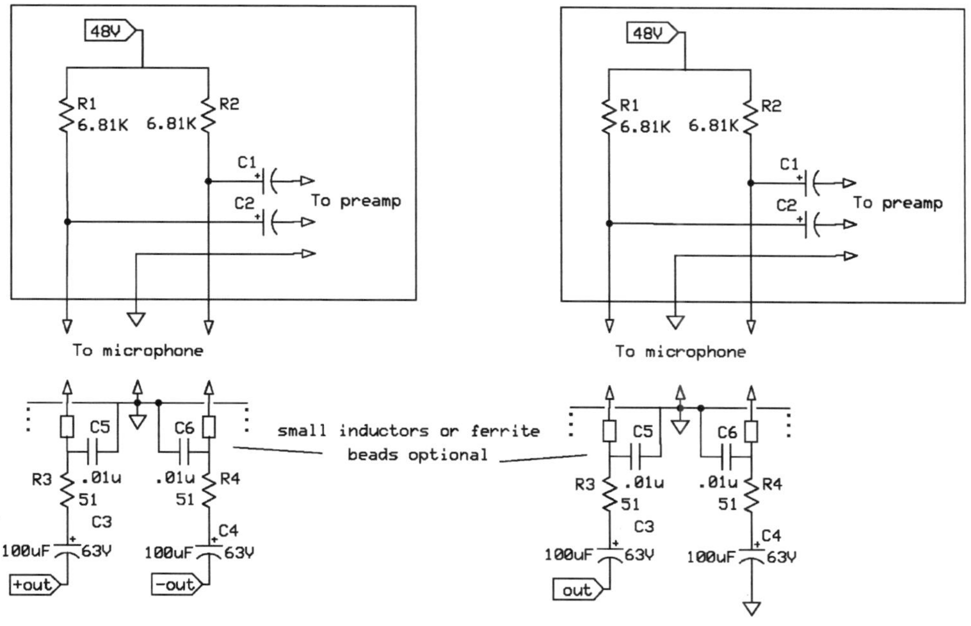

Figure 3: Phantom Power: Left – Balanced; Right Pseudo-Balanced

phone 48V - 3.405 kΩ * I_s, that is, at 14.1 mA total supply current there is no voltage left so some standard is needed for the maximum allowed current draw from the microphone. 10mA is a practical maximum that can be supplied and still leave enough voltage (~14V) for the rest of the circuit. On the other hand this voltage is often used to bias the capsule which can need nearly 40V so it is clear in that case the circuit must draw much less.

Also shown in figure 3 are some recommended components to prevent unwanted external interference. In the pseudo-differential connection termination is provided for common mode interference by these same components. The subjects of grounding and radio-frequency interference can be very involved and there is not room here to fully address them, but in any case it should be noted that the microphone body, cable shield, and preamp case should form a complete solid metallic enclosure with the common (XLR pin 1) bonded immediately at each end.

Two Balanced Transformer-less Microphones

I will first discuss what has become known as the Schoeps balanced transformer-less microphone. This clever circuit is usually attributed to Jörg Wuttke of Schoeps. There are two stages to this circuit, at the input we have the JFET phase splitter that was mentioned in part 1 and this feeds a low output impedance buffer stage via capacitive coupling (**Figure 4**). First I will walk through the design considerations for the phase splitter.

A sensible initial goal is to maximize the linear output range. For the phase splitter we will use equal values for R_1 and R_2 otherwise there will be an undesirable common mode output. The gate current is negligible so it is fair to consider I_d and I_s to be equal and $V_{ds} = V_s - 2RI_d$. The simplest way to make a good first guess at a target for the bias point is to pick a quiescent drain current such that V_{ds} is at ½ of the supply voltage. Taking 1mA as a good value for I_d and considering operation from a 9V battery we have 4.5V/1mA as the value for the total resistance. 2.2kΩ is the closest standard value for R1 and R2. The bias is then adjusted by R_3 for $V_{ds} = V_s/2$. For a mike that is going to see a lot of use/abuse it might be a good idea to replace the pot with two resistors at some point. If you are using a capsule with built-in JFET (modified for three wire operation) simply measure the operating current and pick resistor values to make V_{ds} equal to ½ the supply voltage.

Gate bias return is provided by a large (1-5GΩ) resistor, which was covered in part 1. Theoretically this resistor should be as large as possible, but if one expresses its noise as a current

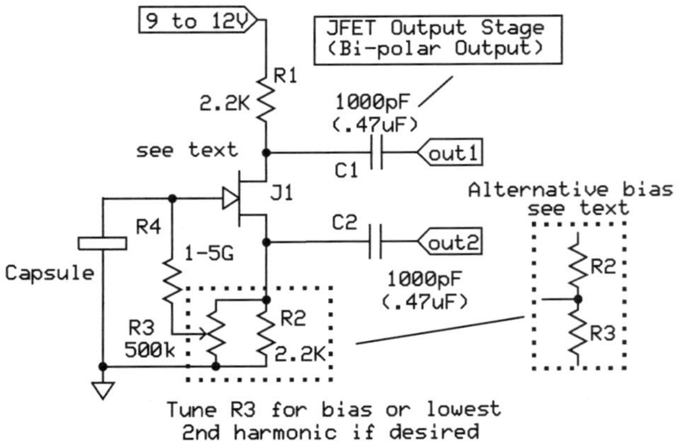

Figure 4: Input Stage – Phase Splitter

($\sqrt{4KT/R}$) and the FET bias current noise as $\sqrt{2qI}$ and equate them, you get I = 2KT/qR. In other words once there is about .052V drop from the bias current through R we are at a diminishing return for noise. That is, for a bias current of 10pA 5GΩ is about the largest practical value for the bias resistor. Most JFETs perform far better than their data sheet specifications, I have typically measured 2-3pA at room temperature, but since this current doubles every 8 degrees K I would still caution those who would make this resistor 20 or 50GΩ. The improvement in noise becomes marginal at best while there is increased opportunity for the bias current to create an undesirable offset voltage. The capsule with the built-in JFET does not need an external bias return resistor.

This ultra-high impedance circuit requires some precautions when built. The easiest method for the DIY'er is to use a Teflon® or equivalent standoff inserted into the PC board to which the JFET's gate, GΩ resistor, and hot lead of the capsule are tied. Even a short piece of #24 wire wrap wire forced into a slightly undersized hole will work.

We can use the analytical solution for output (V_{ds}) vs. input voltage at the gate to obtain approximations for gain and distortion of this circuit. For this we turn again to Leach [4] where his equation (9) with simplification yields an expression for V_{ds} vs. V_{in}. Here I have changed the notation to V_p rather than V_{to} for the pinch off voltage:

$$V_{ds} = 9 - (\frac{1}{\beta R} + 2(V_{in} - V_p) - \frac{\left(1 + 4\beta R(V_{in} - V_p)\right)^{\frac{1}{2}}}{\beta R})$$

where $\beta = The\ JFET\ transconductance\ parameter - \frac{I_{dss}}{V_p^2}$,

$V_p = The\ JFET\ pinchoff\ voltage\ (negative\ for\ n - channel)$

and $R = The\ value\ of\ R1\ or\ R2.$

Since the drain and source voltages are exactly out of phase the ideal gain is 6dB (the 2 in the second term). This would be the case for a very large β, the transconductance parameter. Due to finite transconductance the gain is always less than this and by expanding the equation as a power series one can estimate both the gain and the distortion, dominated by second harmonic. Rather than belabor the algebra we can substitute β and V_p of a real JFET, and in a two step process solve for optimum bias point and estimate the performance. First we must solve for the quiescent V_{in} that gives a DC V_{ds} (output voltage) of 4.5V, then expand the transfer function for AC input signal around that point (using a basic polynomial expansion) and determine the gain and distortion components. First take the J305:

$$J305:\quad \beta = .001, V_p = -2.64V$$

$$V_{ds} = 4.5V\ @\ V_{in(DC)} = 0.62V$$

$$V_{ds} = 4.5V - 1.63 * V_{in(AC)} - 0.027 * V_{in(AC)}^2 + 0.004 * V_{in(AC)}^3 \cdots$$

With this low gm JFET the gain is only 1.63, so let us compare this to a BF862, a high gm JFET, before making any conclusions:

$$BF862: \quad \beta = .035, V_p = -0.65V$$

$$V_{ds} = 4.5V \ @ \ V_{in(DC)} = 1.77V$$

$$V_{ds} = 4.5V - 1.93 * V_{in(AC)} - 0.0076 * V_{in(AC)}^2 + 0.0016 * V_{in(AC)}^3 \cdots$$

Now the gain is 1.93 (almost ideal), but notice the distortion components are also much lower.

R1, R2, and J1 contribute most of the input referred noise. The noise of R1 and R2 combine which is a 3dB increase over the noise of one but then that value is divided by ~2 to get the net contribution which is approximately the noise of 1.1kΩ. This is added to the equivalent noise resistance of the JFET (~2/(3*gm)) . In most cases R1 and R2 dominate the input noise.

In the simple JFET model the characteristic curves in saturation are straight lines so the contribution of the JFETs output conductance is linear. This is not the case for a real JFET; the saturation region is soft near the y axis. Because gm increases as I_d increases the gain goes up with increasing input, while the curvature due to the saturation region of a real JFET is downward (gain decreases with increasing input). If the operating point is adjusted there is a point where these two effects cancel (see part 1) and fortuitously this is near the point that maximizes the total output swing (V_{ds} approaching 0). From the above one can see that a high I_{dss} JFET performs better making the distortion trim less critical and maximizing the gain.

There is a hidden asymmetry in this circuit. The output impedance of a source follower is roughly the inverse of the transconductance (1/gm) so the resistance looking into the source is 1/gm in parallel with R_2. The drain output conductance on the other hand is high so at the drain the impedance is dominated by R_1. This asymmetry affects any errors reflected back from the output so that balanced errors will cancel less well.

The second stage (a differential buffer; **Figure 5**) is where the real cleverness lies. As previously mentioned professional sound equipment is dominated by the use of balanced drive in order to eliminate common mode coupled noise. Many mixing consoles, preamps, and even many portable recorders provide the standard 48V phantom power. Figure 5 is one version of the classic circuit. Two matched PNP transistors are diode connected with large resistors and act as diodes to the DC phantom power current but as emitter followers to the signal which is AC coupled to the bases.Their currents are joined and reused to power up the rest of the circuit. In normal operation the outputs are push pull and out of phase so the total output current is constant. If the rest of the circuitry operates near class A the total current in the microphone is also nearly constant making this a perfect place for shunt regulation. A shunt regulator can be as simple as a zener diode and bypass capacitor. Care must be taken to insure that the average DC current demanded by the rest of the circuit does not starve the shunt regulator. Several manufacturers now make IC shunt regulators that are much more forgiving

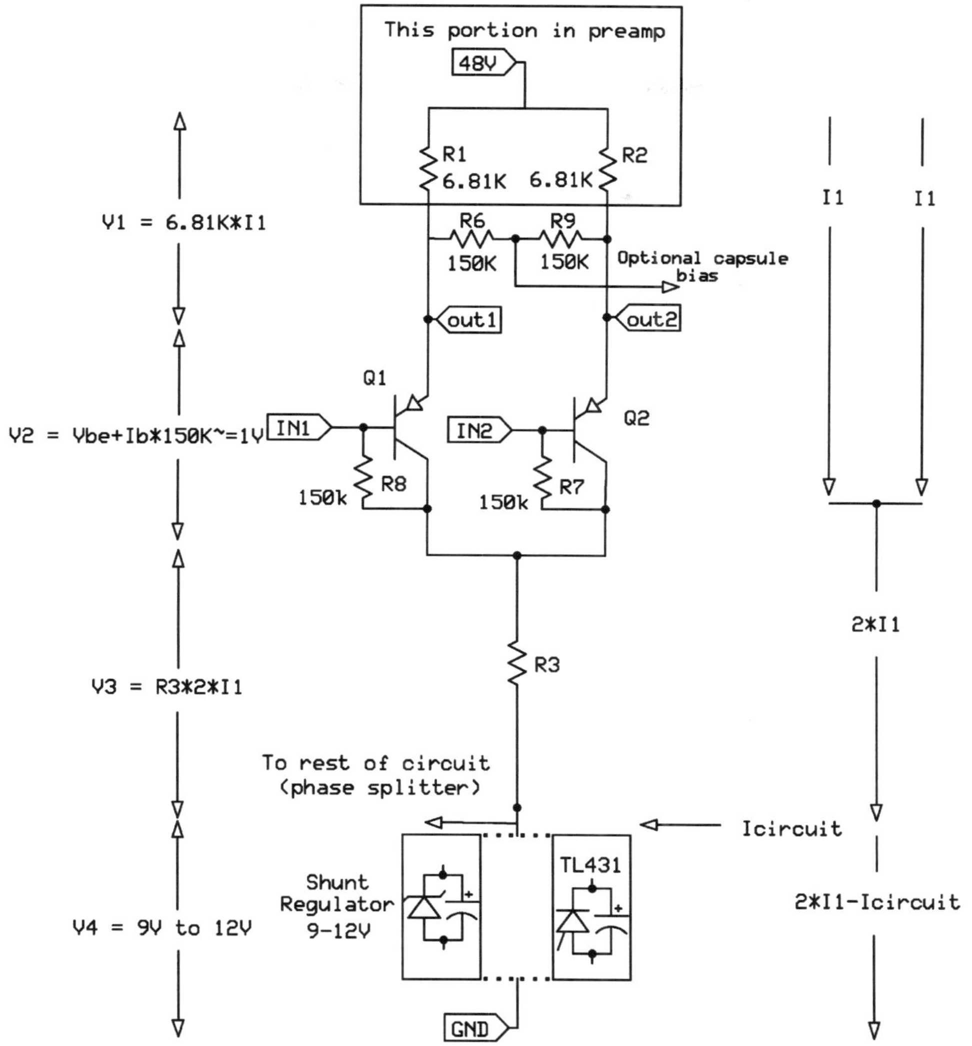

Figure 5: Bipolar Version of Output Stage

on the amount of "keep alive" current; the TL431 is popular in this socket. In figure 5 on the right are listed the currents in each branch of the circuit while on the left the voltage drop across each part of the circuit is listed. It is clear that once the operating current (I1) and shunt regulator voltage are chosen the value for R3 (the last unknown) can be computed. Note that the drop across the diode connected transistors can be taken as ~1V in all cases since V_{be} and the drop across the 150kΩ resistors does not have a lot of variance.

There are some subtle issues with this approach that can be problematic. The V_{ce} of each transistor

is in part derived from the drop of the base current in the 150K resistors, which means betas should be matched. If beta is very high the maximum output swing could be compromised, remember for a differential output the total current remains the same so the point at the junction of the two collectors remains relatively constant while the bases are swinging the full output voltage. If beta is low base current errors reflected into the output resistance of the first stage can cause both noise and distortion. The best choice here is a low noise pnp with beta around 400 and a low V_{cesat}, the 2N5087, or BC859C are examples. A zener diode (4.7V) should be connected, cathode to emitter and anode to collector, of these transistors. This will prevent reverse breakdown of the base emitter junctions of these transistors which could permanently degrade β.
Now we can compare the gain and noise performance estimates vs. a SPICE simulation, using figure 4 and figure 5 using a BF862 (gm = 12mS@1mA) and pnp's with β = 400. Note, the noise contributions at the source of the JFET are slightly reduced by the fact that the resistance looking into the source is not zero but ~1/gm which gives ~4% reduction (~ (1/gm)/2.2kΩ) to errors injected into the source.

$$1.\,Gain$$

$$Estimate - \ 5.7dB; simulation: \ -4.86dB \ \ (.9dB\ lost\ due\ to\ input\ C)$$

$$2.\,Noise\ of\ R1\ (figure\ 4) - \frac{\sqrt{4KTR}}{Gain}$$

$$Estimate - 3.1nV/\sqrt{Hz}; simulation: \ -3.0nV/\sqrt{Hz}$$

$$3.\,Noise\ of\ R2\ (figure\ 4) - \frac{\sqrt{4KTR}}{Gain}$$

$$Estimate: \frac{3.1nV}{\sqrt{Hz}}; \ Simulation: \ 2.9nV/\sqrt{Hz} \ \ (due\ to\ JFET\ source)$$

$$4.\,Noise\ of\ Q1\ and\ Q2\ (figure\ 5) - \frac{R_{1\ or\ 2\ (fig.4)}\sqrt{2*2qI_b}}{Gain} \ \ (only\ I_b\ contribute$$

$$Estimate: \frac{1.8nV}{\sqrt{Hz}}; \ Simulation: \ 1.9nV/\sqrt{Hz}$$

$$5.\,Noise\ of\ J1\ (figure\ 4): \sqrt{{}^{8KT}/_{3gm}}$$

$$Estimate: \frac{1.0nV}{\sqrt{Hz}}; \ Simulation: \ 1.1nV/\sqrt{Hz}$$

The results agree quite closely and it is worth noting that the JFET is only a minor noise contributor and furthermore substituting a "low noise" 2SK170 will decrease the gain by an additional 2dB (to only 2.8dB) with its high input capacitance and effectively increase the input referred noise.

I now want to explore the benefits of replacing the PNP transistors with P-channel JFETs. This can have several advantages not the least of which is eliminating all issues dealing with the base current. Unfortunately there are very few choices for this JFET, like the still available but getting harder to find 2SJ74 and its replacement the LSJ74 from Linear Systems. The 2SJ107 is also still available and should work quite well. Interfet Corporation lists a replacement for the NEC 2SJ44 (IFP44) but there is insufficient information on the datasheet to determine its suitability. There are other choices of NOS, new old stock of discontinued devices, but they are often very expensive and hard to find. I will use the L/2SJ74 since it appears that it will remain available for the foreseeable future.

I will first discuss a 9V battery powered version (**figure 6**). "diode" connected JFETs are not diodes but appear as current sources to the DC bias and as source followers to the signal. With FETs the bias resistor can be much larger than with bi-polar devices since the gate current is usually pico-amps.

For the battery powered version, since we have access to the supply directly, I use an LED of appropriate color running at a small amount of current to generate a bias voltage with reference to the positive supply. The LED voltage plus the V_{gs} of the 2SJ74's should total around 2.5V but this does not need to be precise. The goal is to set the operating current of the JFETs at around 1.0-1.5 mA each. Done this way the full load resistor (R4, R5) degenerates the source follower and any JFET mismatch in V_{gs} works against the drop across R4 or R5 which is >2V. For a 2SJ74Bl the max offset for randomly chosen devices is 0.17V which in turn gives a maximum current difference of 85µA and more importantly the second harmonic due to mismatch is 25 times lower than the distortion of the phase splitter itself. In essence we have designed around the need to match devices. The drains of J2 and J3 go directly to ground and now the output can swing several volts peak to peak without clipping, another advantage. A simpler solution, inset in figure 6, requires resistor selection and matching to achieve the same result.

At the input we now can use very large bias resistors and smaller high quality film capacitors for AC coupling. The capacitor value is not determined by the time constant but rather it needs to be large enough to roll off the noise of the gate resistors. Large value resistors are readily available but the series combination of ordinary 22Meg resistors is very cost effective. I would use 2 or 3 in series to make either a 44Meg or 66Meg value for R6 and R7. The coupling capacitor

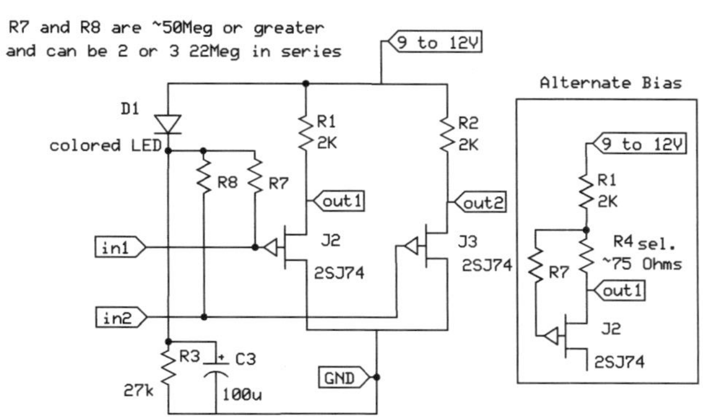

Figure 6: JFET Version of Output Stage for Battery Power

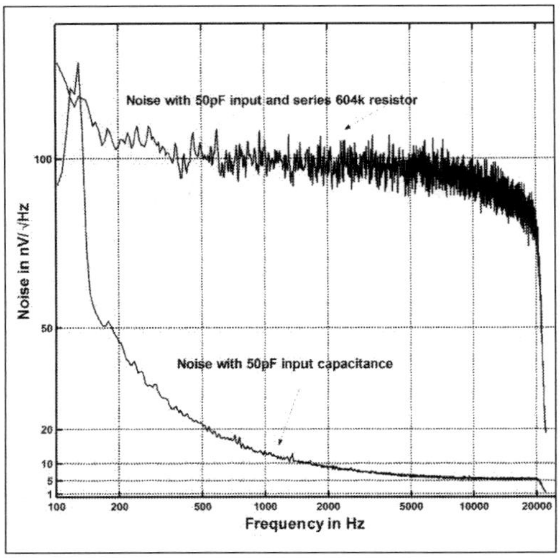

Figure 7: Electrical Noise of Complete Circuit with Calibration

should be large enough so that the noise of R6 and R7 referred to the input is less than the contribution of the capsule bias resistor. 1000 – 2000pF is enough in most cases so one can use polystyrene or another high quality film capacitor at nominal expense.

Two plots summarize the performance. **Figure 7** is a plot of the noise and **figure 8** is the response to a 1kHz signal capacitively coupled into the input to produce 1V peak to peak at the output and with the bias trimmed for minimum second harmonic [1]. The dominant third harmonic is -88dB and only increases to -72 dB at 0dB out which is only 0.025% THD at 2V p-p so the net THD will likely be dominated by the capsule and not the electronics.

The current noise of the pnp transistors has been eliminated and the 2SJ74's add only about half of that back for an estimated noise of 4.5nV√Hz. From figure 7 we can see that all non-idealities including signal attenuation due to input capacitance raise that to only ~5nV√Hz.

The phantom powered version of this circuit must overcome a fundamental roadblock: there is no direct access to the supply voltage so the gate bias, which now will be at a higher potential than any other point in the circuit, must be separately generated. In order to achieve this we must call into play an unusual component. Solid state relays need a completely isolated trigger voltage, which is generated by a photovoltaic cell coupled to an LED, to make a 7.5 to 8V (at no load) floating voltage source. The currents we need are at the pico-amp level (gate currents) so this is a perfect application. There are multiple sources for this component [5]; one of the lowest cost and most readily available is the Panasonic APV1121S which is in a small 4 lead package. In general ignore the datasheet specifications on these parts since we are using them well outside the intended applications. All parts that I have tried produce nearly their full output (unloaded) at <1mA of LED current.

Figure 9 shows everything falling into place. A JFET connected as a current source is placed in series and rather than floating up to a point de-

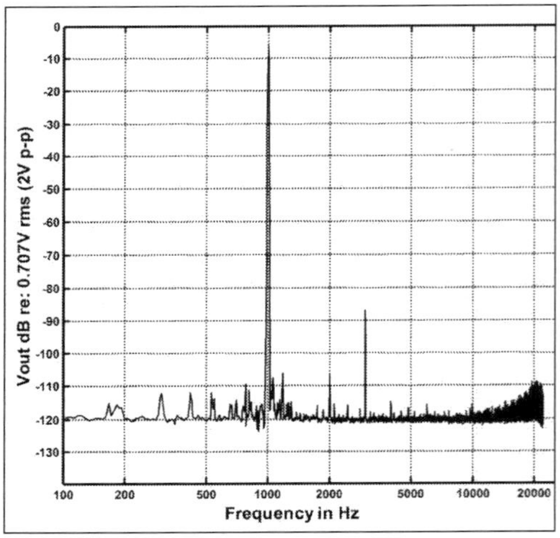

Figure 8: Distortion at 1VPeak tot Peak Output with Distortion Trim

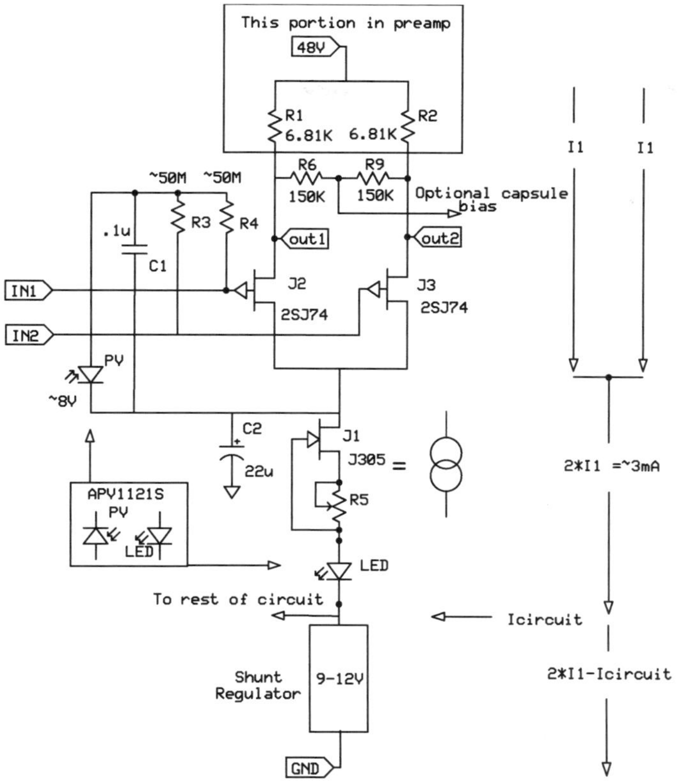

Figure 9: JFET Version of Output Stage for Phantom Power

termined by voltage drops the circuit floats up to a point where the combination of J2 and J3's current equals the current source (J4), with the junction of these three current sources acting as a high impedance (gain) node with feedback to the JFET gates, an exact analog to the self biasing connection for a bipolar transistor. This loop is dominant pole compensated by C2 and because it is a common-mode loop we take the time constant from R1||R2 and C2. The input coupling time constant formed by R3 or R4 and the coupling capacitors from the phase-splitter cause excess phase shift inside this loop so the main time constant should be chosen to be at least 1.5X this to achieve a well damped turn-on transient. For the values here, 50MΩ for R3 and R4 and 1000pF input coupling capacitance and 3.4kΩ and 22µF for the main loop the settling is well behaved (~1sec). Below the current source there is the shunt regulator for the input stage that reuses the bias current just as the previous circuit. The first stage and AC coupling to the output remain the same as in the battery powered version.

Op-amps in Microphone Circuits

In 2004 I floated the idea of bootstrapping a simple all in one capsule with a complimentary pair of emitter followers. The result (**Figure 10**) had an impressive dynamic range with low distortion as well as low output impedance.

The same idea can be accomplished using a single low power rail to rail op-amp, **figure 11**. The JFET is biased in a fashion similar to the phase splitter but the components are chosen with a very different goal. Here an op-amp drives the drain and the bottom of the gate biasing network via C1 and C2 bootstrapping everything to follow the input voltage. With the bootstrapping the drain current of the JFET is signal independent so it acts like an almost perfect source follower, even the effects of drain to gate capacitance (except for noise) are reduced to negligible proportions.

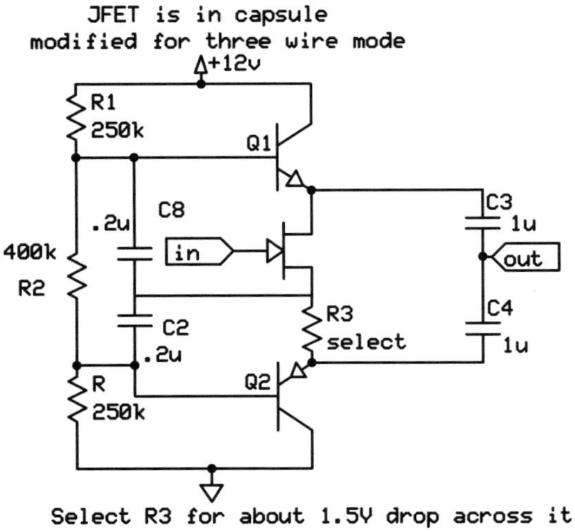

Figure 10: A Fully Bootstrapped Capsule Microphone

The resistors are chosen with two considerations. The op-amp must drive R3 and R4 so their value should be maximized within the constraints of JFET quiescent current and drain to source voltage. Since in this circuit the drain to source voltage is forced constant we simply pick a DC value. The total drop in the source network should leave the source at about mid-supply so that output swing is maximized. Now for some sensible starting values, 3V drain to source voltage and 1mA drain current are good. This would give approximately 4K for the total of R1+R2+R3 and 2K for R4. R2 and R3 can again be a 2KΩ resistor paralleled with a potentiometer that is adjusted for the desired bias or two resistors in the same ratio. If using a three-wire capsule just select R1 to put the source at mid-scale and R4 to get the drain to source voltage down to 2-3V.

The bootstrapping removes the effect of C_{gd} but the noise of the op-amp still appears at the drain, and with low capacitance capsules this can appear at the input via capacitive division so I would rec-

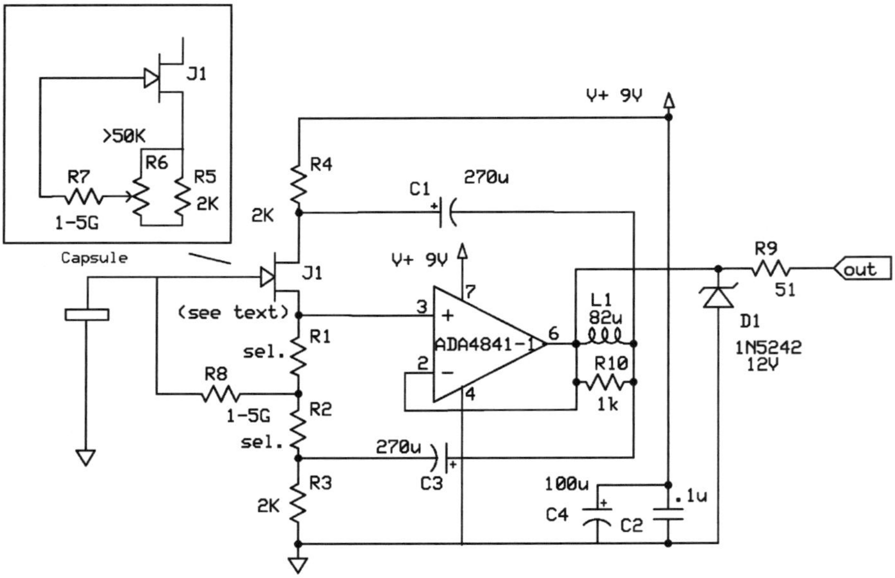

Figure 11: Op-amp Bootstrapped Source Follower with Rail to Rail Output

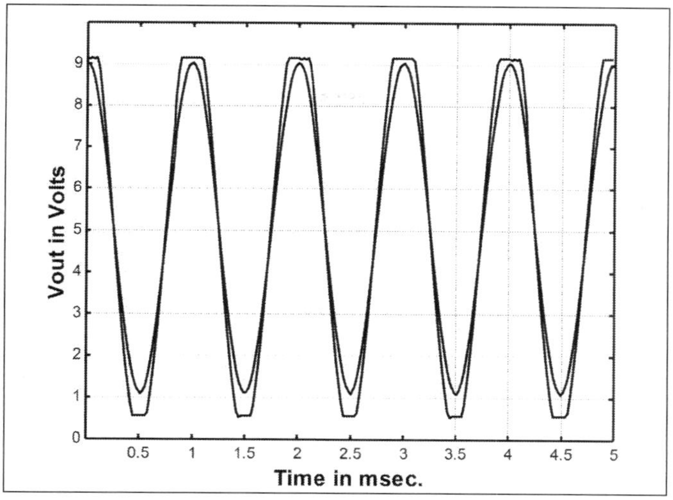

Figure 12: Output of Circuit in Figure 11 on a 9V Battert

ommend lower capacitance JFETs such as the 2SK222, BF862, J305/9/10, 2N4416 etc. The pinch-off voltage of the JFET should be low enough ($|V_{gs}| < -2V$ at 1mA) so that it is possible to adjust the bias point. Some J305/9/10's and 2N4416's might go over this number at their I_{dss} extremes. Also shown is a snubbing network, L1 and R10, which prevents high frequency oscillation.

Anytime when there is danger of "hot-plugging" an op-amp into a phantom powered preamp there should be a zener diode clamp selected to be just over the designed supply voltage, this is shown in figure 11 but applies to all of the op-amp circuits.

Figure 12 shows that, yes, you can get an output peak to peak value very close to the full supply voltage in this configuration, even though it starts as a simple source follower. The plot shows an undistorted 8V p-p out superimposed on the output in clipping.

The Charge Amplifier

The charge amplifier was invented in 1950 by William Kistler. A high gain amplifier (op-amp), **Figure 13**, in the inverting configuration is used with capacitors as feedback elements. Conservation of charge forced by the op-amp does all the work (Vo = (C1/C2)*Vin). The feedback resistor is used to provide DC return for the input bias current, so as before it wants to be as large as possible and the amplifier should have very low input current. When I first built some experimental microphones using a charge amplifier I didn't realize that the Neumann U47-FET was based on the same technique. So even though it is uncommon in the world of high end microphones, at least one of the more prestigious manufacturers offers a microphone based on this type of amplifier.

At this point it is necessary to digress a bit and discuss at least one performance aspect where the charge amplifier is at a disadvantage. There are two modes of operation at play here. Recalling the basic relationship between charge and voltage on a capacitor, Q = C*V, we see that either Q or V can be held constant while the sound pressure varies C. All previous circuits operated in the constant charge mode where the voltage on the capsule is measured with a very high impedance amplifier.

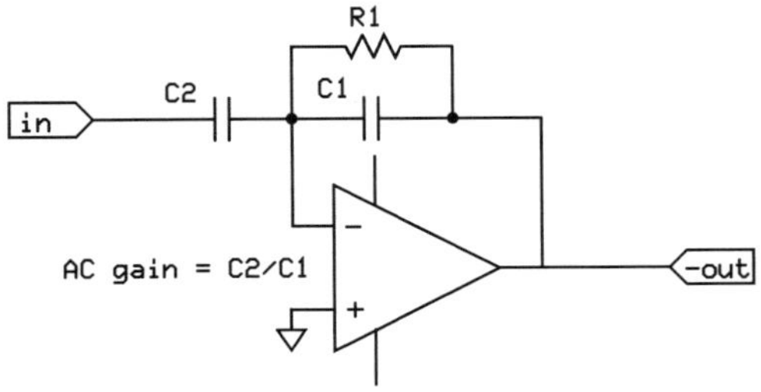

Figure 13: Basic Charge Amplifier Concept

Fredricksen [6] showed that in this mode the sound pressure to voltage out is (to a first order) a linear function. Borrowing his terminology, C = Co/(1+y), where Co is the capsule capacitance at rest and y is the relative displacement of the diaphragm (equal to the displacement of the air). So when we take V = Q/C the relative displacement is in the numerator, i.e. the voltage out is linearly proportional to it. For example at polarizing voltage of 50V when the acoustic pressure changes the capsule capacitance +-2% the output is +-1V (a lot actually, at 20mV/Pa that's about 125dB SPL).

The charge amplifier operates in the constant voltage mode and there is inherent non-linearity in the transfer function. Now the relative displacement is in the denominator and the charge transfer is proportional to $1/(1+y) = 1-y+y^2\ldots$, and at the same 2% displacement the output is +1.02 to -.98 or ~1% second harmonic distortion. This should only be a problem at very high SPL, but we should be aware of it.

Figure 14 shows a composite amplifier that mates a single ended JFET input with a rail to rail low power bi-polar op-amp. For noise to quiescent current ratio it is hard to beat a bi-polar input amplifier (here 1.7nV/√Hz at 1.5mA I_{supply}) and with the single ended JFET input the total noise is minimized. A prototype was built with the Primo EM21/23 style capsule and a gain of 11 using a 5pF feedback capacitor ($C_{capsule}$ = 55pF).

With this configuration we can have virtually everything one would want for a low cost microphone. Gain is easily set simply with a capacitor; this could even be a small dip switched set of gain setting capacitors. The output swings almost rail to rail, so even set as it is here at 110mV/Pa (-19dB re 1Pa) there is a large headroom. This means with many field recorders we can get the low noise and wide

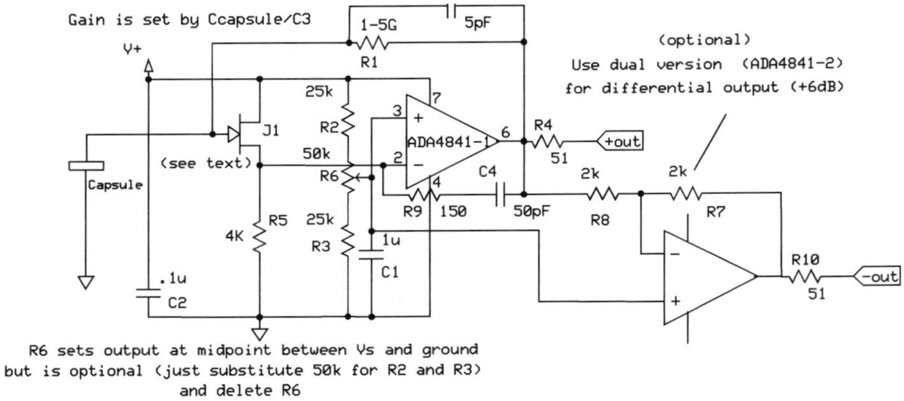

Figure 14: Op-amp Based Charge Amplifier

dynamic range performance without the use of an external pre-amp.
The TSB-165A ($7.50 from JLI electronics) is easily adapted to this circuit simply with different values for the feedback capacitance.There are several choices for the op-amps, the builder should be careful that the noise, supply current and audio performance (bandwidth and slew rate) are adequate.
Matching prototypes of this circuit were used to make a live recording, a sample of which we have kindly been given permission to put up on the Linear Audio web site [10]. The recording rig is shown in **figure 15**, the two microphones are mounted in what is called a head-spaced barrier [7] where the microphones are set back ~0.5" from the front edge. This is a quasi-binaural technique which some think is a good compromise for either headphone or loudspeaker listening. The microphones are held in place by two modified hair curlers and if necessary a windscreen (modified car wash mittens) is placed over each side of the assembly.

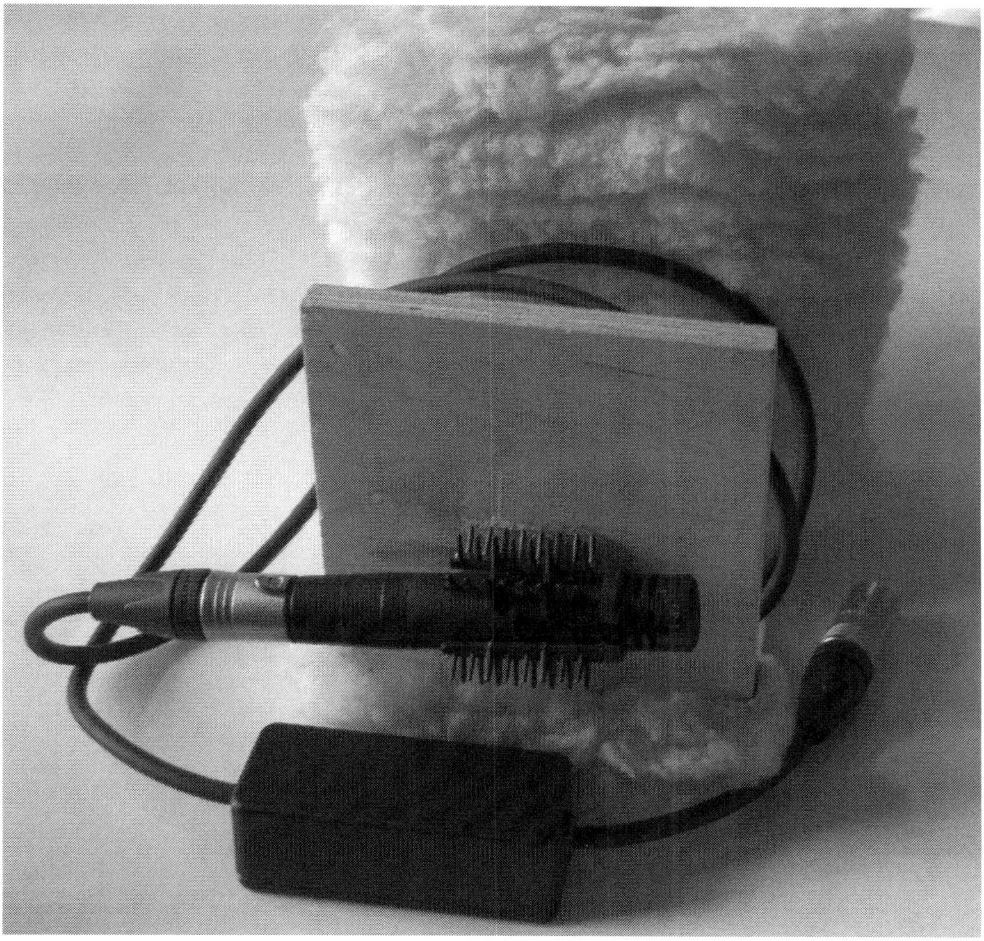

Figure 15: A Head Spaced Barrier Recording Setup

A few notes on the recording – This was an impromptu living room recording, with no acoustical preparation of any surface, no sound check, etc. The microphone assembly (using the EM-21 cardioid pattern capsules) was simply placed in the front center seat and the recording was done in one long session on a Fostex FR -2LE digital field recorder at 24/96 resolution.

The Capacitive Feedback Amplifier

The last op-amp circuit that I will present takes the charge amplifier concept and flips it so that the capsule is again used in the constant charge mode (**figure 16**). The feedback is taken from the output of the op-amp via C3 and C4 and the closed loop gain is simply, (C4+C3)/C3. Small resistors R7 and R8 isolate the op-amp from the capacitive load and prevent high frequency problems. The astute reader might notice that frequency shaping could be added to the feedback network, this level of complexity is beyond this article but a similar (but not quite the same) circuit is discussed in the literature [8].

This configuration is one where the capsule is not connected directly to ground and is a good retrofit for the popular clones of the classic Neumann "bottle" microphones. These are available very economically from numerous sources and there is room in the body for a circuit board and even the battery if desired. The external capsule bias can be added as shown with minimal impact on noise performance (see External Capsule Bias section below).

Balanced vs. Pseudo-balanced Output

As shown in figure 3 both single ended and balanced outputs can be used with standard XLR cables by providing balanced impedance to interfering signals. With care in grounding and shielding the single ended circuits can work quite well especially over short distances. Fully balanced outputs are easily added to these op-amp circuits by using a dual op-amp with the second half either config-

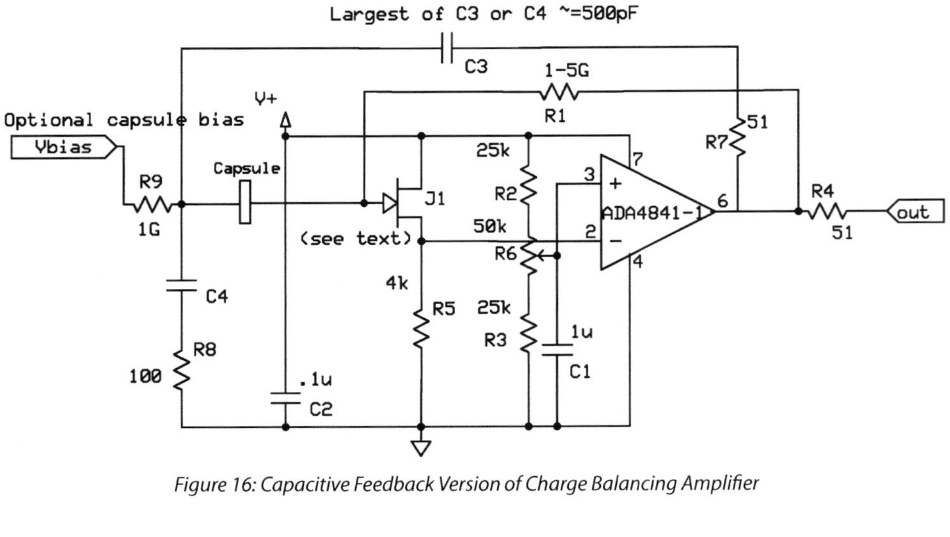

Figure 16: Capacitive Feedback Version of Charge Balancing Amplifier

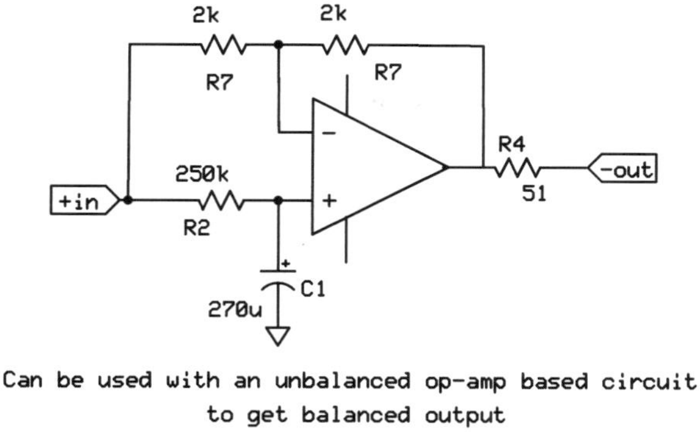

Figure 17: Self Biasing Inverter for Adding Second Phase to Output

ured as a self-biasing inverter, **figure 17**, or if a $V_S/2$ reference is available it simply can be used for both op-amps. It is very important to realize that these circuits operate in class A/B rather than class A, which means that the supply current will increase at higher output drive levels. It is very important to not have input dependent current demand which would cause signal dependent common mode signals on our balanced lines when using a phantom power connection, so the shunt regulator should be set to the highest current that could be demanded by the circuit. The circuit in figure 14, when configured for differential output, demands <9mA when driving a full 16V p-p into a 2k load so it can fit into the 10 mA maximum budget. If one wants to drive a preamp with a 200 Ohm input impedance a 2K: 200 transformer would be necessary.

External Capsule Bias

To incorporate an external capsule bias there are several options available, the usual solution being a small Colpitts oscillator operating well outside the audio range with a Cockcroft-Walton voltage multiplier providing boost up to the designed voltage. More voltage than the recommended value can stress the diaphragm outside of its intended operating range, eventually shorting it which is potentially fatal to the diaphragm. Various well tested versions of these circuits are documented in several places notably the files section of the Yahoo group micbuilders.

CMOS voltage multipliers are also used, the simplest is a 1044 charge pump IC [9].The best version I found is the Microchip TC1044 which has the features of up to 12V operation and provision for putting the frequency of the pump just outside the audio range. The cheapest is probably a single hex inverter (CD40106B), **figure 18**; here the frequency can be placed well above the audio range (the only penalty is that CMOS uses more current at higher clock rates). The current draw is essentially zero once the capsule is charged so the current draw is simply the wasted power in switching the CMOS gate capacitances. The entire circuit can be mounted in a small isolated enclosure to insure there is no leakage of high frequency energy.

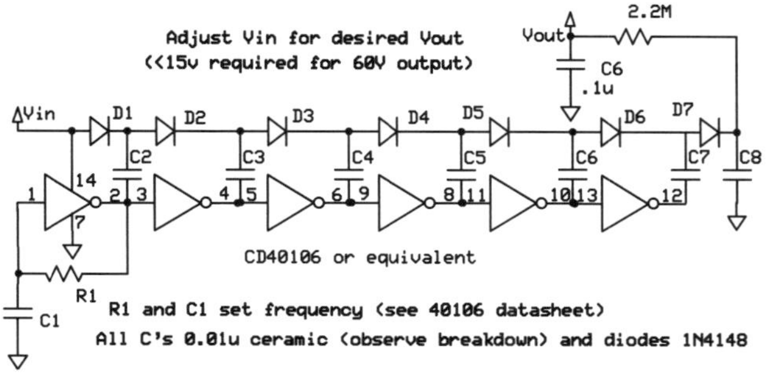

Figure 18: Low Cost CMOS Capsule Bias Generator

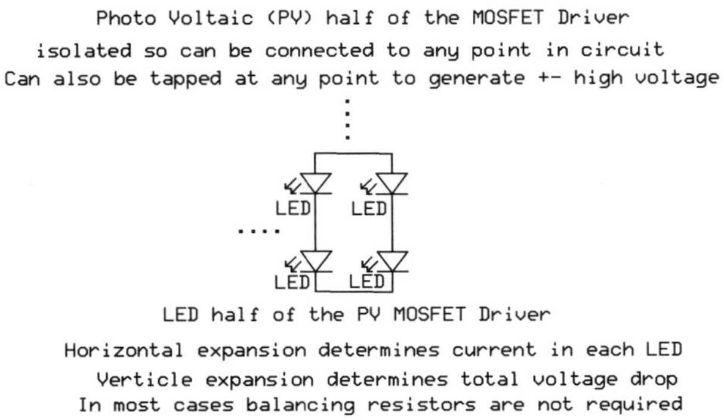

Figure 19: Various Connections for Photo Voltaic Bias Generators

I would like to introduce a third option and returnto the photovoltaic MOSFET driver. A survey of available parts turns up numerous examples some with up to 36V in one package. Not all of these devices are readily available to the DIY community but for completeness I listed as many as I could locate [5]. They all share a common feature: each cell (2 in a dual) has a single LED that controls the output (~7.5V – 18V). As previously stated I am using these devices well outside their intended application, and since we need virtually no current, we have useful output as soon as the LED is activated. For the Panasonic APV1121S, ~1mA LED current produces ~ 8V. The great feature is that the PV voltage is completely galvanically isolated so the cells can be stacked and referenced to any circuit voltage, for instance we could simply boost the highest voltage remaining on a phantom power circuit to a higher voltage, or we could ground the center tap on a series of these and get a dual polarity high voltage bias for a multi-pattern capsule. Another feature is that the driving LED can be series/paralleled in any combination to optimally use the current/voltage available, which means in most applications the current that biases the circuit can be reused to generate the capsule bias with no penalty and no high frequency energy at all. **Figure 19** shows a few suggestions for different connections. There is an additional advantage to applying external bias directly to the capsule body, **figure 20**, in that the noise of the second bias resistor is eliminated. This is most easily accomplished with capsules that are designed to be enclosed in a fully shielded microphone body such as the copies of the classic Neumann capsules.

This feature can be worked into any of the circuits here if care is taken to not let the LED voltages subtract from headroom requirements and that the LEDs have sufficient current. In some cases the higher output PV MOSFET drivers would be more appropriate.

I will use figure 6 to demonstrate the addition of external bias with these PV arrays. 6 PV arrays are used and this voltage is added to the battery voltage (**Figure 21**) so that the entire circuit still operates on ~5mA and 9V. In figure 5 we would place the LEDs in series with

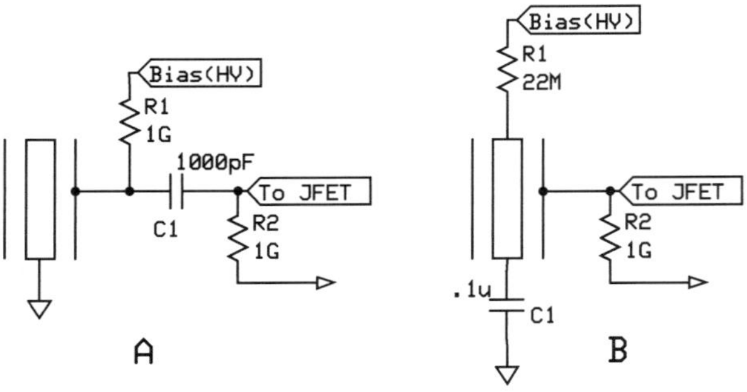

Figure 20: Two Ways to Connect External Bias to Capsule

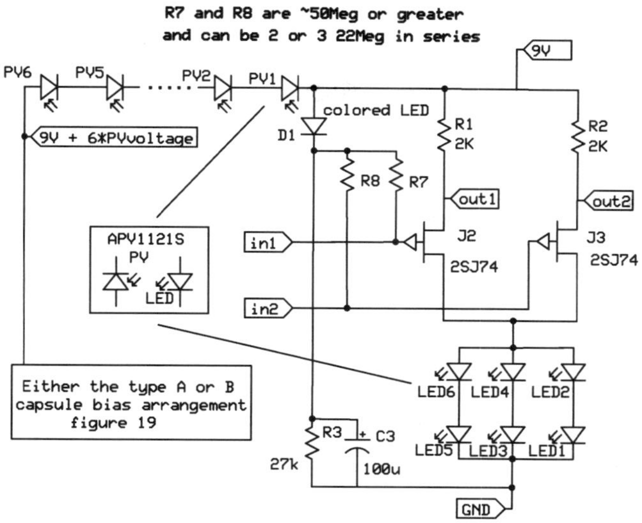

Figure 21: Adding Photo Voltaic Bias to Circuit of Figure 6

R3 and re-compute R3's value taking into account the voltage drop across the LEDs, which depends on the particular device (for the Panasonic APV1121S it's about 1.2V). This makes for reuse of all of the supply current, and in figure 9 we have one APV1121S already so if the remaining voltage at point A is 36V we need only 3 more to reach 60V. These 4 LEDs could be simply series paralleled to give each LED >1mA current and lose only ~1.2V extra from the stack.

For the Adventuresome Builder

One last circuit trick that you can apply to some of the circuits here offers the opportunity of eliminating the GΩ resistor which can be hard to find. The circuit of figure 11 is the easiest candidate since the bootstrap voltage needed is already provided. The trick is to use two back to back low leakage diodes as a substitute for the high valued resistor and bootstrap the voltage across the diode to fol-

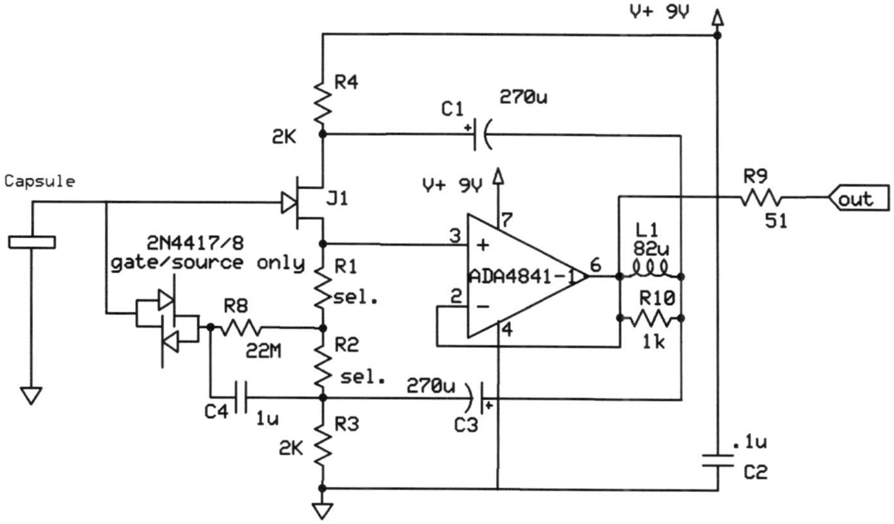

Figure 22: Substituting Low Leakage Diodes for GΩ Resistor

low the input signal so that they remain at their zero crossing resistance which can be extremely large ($\sim V_t/I_{sat}$) where V_t is the thermal voltage ~0.0259V at 300K and I_{sat} can be in the femto-amps for the right junction. A perfect choice is the gate to source junction of a JFET, the 2N4417/8 JFET used in the ionization chamber of smoke detectors is a good example. **Figure 22** shows how this would be implemented in the circuit of figure 11. In essence one or other of the diodes will forward bias just enough to balance the net leakage current at the JFET gate which gives a resistance of V_t divided by this current which for 10pA is ~2.5GΩ. The reason for needing low leakage diodes is that this effective resistance can never be less than V_t/I_{sat}.

Acknowledgements

My approach in this article has emphasized the circuit design aspect of the recording microphone, essentially the problems associated with the high impedance capacitive transducer. I have never worked in the audio industry per se which is one reason I recommended the technical library of B&K as a reference. One could say it speaks my language: that of a scientific instrument maker. There is however a whole other facet to this subject, the microphone as the recording engineer's "instrument". The great designers such as Georg Neumann, Stephan Peus, and Jörg Wuttke have worked inside the audio industry and had access to experience that I could never have. This work certainly owes them and others a great debt. There is a vast amount of free literature available that covers both the technical and artistic aspects of microphones and the recording process, one example being "Microphones" by Bore and Peus available on the Neumann web site.

I would also like to thank some people who have contributed to my knowledge on this subject. From the Yahoo micbuilders group, Jim Gillespie, Eric Benjamin, Richard Lee, and Mike Rooke among others; Jim especially for actually trying out my hair-brained ideas and making them available to the community at large. Recording engineers Martin Kantola and Goran Finnberg for many helpful comments as well as Rob Danielson, (retired) professor of recording arts. Finally, Dimitri Danyuk for reviewing several of these circuits. At some level this article is the sum of a lot of ideas that have been floating around with a few new additions.

References

[1] Wurcer, Scott, "Low noise Microphone Preamplifiers, part I", Linear Audio Volume 1, April 2011, pp. 97.

[2] http://www.jlielectronics.com/

[3] Primo Microphones, Inc., 1805 Couch Drive, McKinney, TX 75069
Phone: 972-548-9807
Fax: 972-548-1351

[4] http://users.ece.gatech.edu/mleach/ece3050/notes/jfet/thejfet.pdf

[5] Panasonic APV1121 single 8V
Vishay VO1263 dual 14.6V
AvagoASSR-V62x dual 7V
DionicsDIG-22-8-30-DDx dual 7V

Semitronics CT222ADMdual 18V
IRC PVI5033R dual 7V
[6] Microphone System for Extremely Low Sound Levels, Erling Frederiksen
http://www.bksv.com
[7] http://www.trackseventeen.com/mic_rigs.html
[8] A Low-Noise High-Output Capacitor Microphone System
JAES Volume 25 Issue 5 pp. 278-283; May 1977, Burwen, Richard S.
[9] http://www.geofex.com/circuits/+9_to_33.htm
[10] A live recorded song using the microphones described in this article can be downloaded at www.linearaudio.net/online resources/online articles . This song is © Southpaw Jones (www.southpawjones.com), and is posted by kind permission. Recording took place at October 2011 at Stuart Yaniger's residence (www.syclotron.com). A separate file of the applause, uncompressed, can be downloaded as well.

Appendix A

Many sound cards do not have low noise pre-amps built in. To facilitate noise measurements a small battery powered preamp can be very useful. The circuit in figure A1 fits the bill. Its high gain will overcome the noise of the input circuits on most sound cards, while in some ways outperforming typical low noise bipolar preamps.

The circuit is another take on the JFET/op-amp composite amplifier, here in a single ended version. For low noise the feedback network uses very low resistances, usually requiring a very large electrolytic capacitor for DC biasing. The approach here takes a second 2SK170 connected as a very low parasitic resistance diode, to raise the source voltage enough so the FET can now be biased from a single supply. The AC coupling is now moved to the input of the op-amp where the coupling capacitor works against 10MΩ, while the gain network is directly connected to ground. The DC bias loop has only this time constant so there is no chance for peaking at the low frequency cutoff.

The 2SK170 has modest gain in this configuration but it is enough so that a generic JFET op-amp will work in U1's socket. With an AD711 1.4nV√Hz was achieved at G = 101, essentially limited by the JFET, diode, and R3.

Setup is simply to trim R6 for 2mA in J1 (4V across R5) and possibly select C3 for the op-amp chosen, just lower R3 for higher gain.

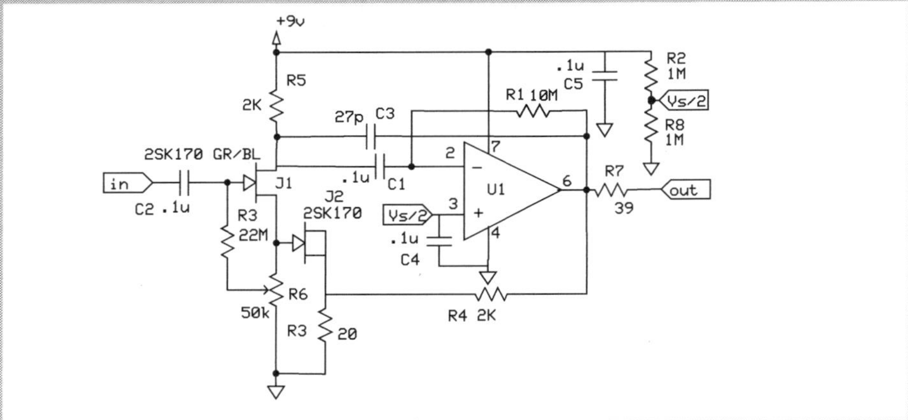

Figure A1: a simple, low-noise mic preamplifier for sound cards

Correcting Transducer Response with an Inverse Resonance Filter

A biquad filter based on the Sallen-Key filter topology

Steven van Raalte

1 Introduction

Over thirty years ago Hans van Maanen published two articles in the Dutch magazine Radio Electronica on the compensation of the mechanical resonance of phono cartridges [1], [2]. In these articles Hans van Maanen explained how the stylus and cantilever mechanical resonance of a phono cartridge behaves as a severely under-damped second order low pass filter. This mechanical filter creates a serious resonance peak around 20 kHz, but it does so by design. The resonance peak compensates for the high frequency loss of the electrical low pass filter that is formed by the cartridge coil and its load impedance. For moving coil cartridges the inductance of the coil is low and the associated low pass filter has its corner frequency outside the audio band, but for moving magnet cartridges this is a real issue. In his articles Hans explained that, despite the virtually flat frequency response of these cartridges up to 20 kHz, the mechanical resonance causes time smearing of transient signals as a result of the severe phase shift in the highest octave of the audio band. He then provided a solution in the form of a resonance compensation filter to 'undo' the mechanical resonance. His filter is based on a normal active second order low pass filter that imitates the mechanical filter, which is incorporated into the feedback path of an additional amplifier. This combination becomes an inverse second order low pass filter, a filter that dips at the resonant frequency and boosts frequencies above the resonant frequency with 12 dB/oct. The problem with this topology is that the phase shift of the low pass filter in the feedback path quickly approaches 180 degrees, eventually turning the amplifier into an oscillator. Hans solved this by the addition of two zeroes in the low pass filter, turning the low pass filter characteristic flat again above a certain frequency, and bending the phase response back to 0 degrees before the critical 180 degrees is reached. Although this works, the stability of the circuit remains marginal.

In 1981, having read the articles of Hans van Maanen, a friend of mine and I were looking for an alternative filter topology to design an inverse resonance filter. We had tried the same approach as Hans had followed and were not happy with the marginal stability of our circuit. It was opamp based and different from Hans' circuit, but the basic idea was the same. We then thought about using a state variable biquad filter. These filters are based on analog computer technology and typically are implemented with two integrators to create different state outputs. By applying feedback and signal summation any type of second order filter can be implemented, with complex poles as well as complex zeroes. In the literature (and on the web) you can find a number of topologies, named after their inventors, like the Tow-Thomas biquad filter and the Kerwin-Huelsman-Newcomb state variable filter. A big advantage of these filters is the possibility to tune the filter characteristics independently, and so these filters are found for instance in the fully parametric equalizer sections of the bigger audio mixing desks. But they need at least three and often four opamps and we wanted something smaller.

For DIY audio purposes it is often fun to try to get the maximum of performance from a minimum number of components. So, we decided to take the minimalist route and not use a state variable filter, but only a unity-gain buffer as the active element. This allows the use of opamps to create a fairly perfect buffer, a fancy discrete BJT or FET buffer, or even a tube cathode follower stage! The filter topology for this is the VCVS (Voltage Controlled Voltage Source) topology, better known as the Sallen-Key topology. Almost all literature, however, only show the standard low pass, high pass and band pass filters for this topology; all filters with complex poles in the transfer function, but no complex zeroes. We then got lucky; our school library received a new book, *Netzwerksynthese in Beispielen* [3]. There we found an exercise that made clear that it is possible to create complex zeroes with the Sallen-Key filter topology. We worked through the math of this filter and designed an inverse resonance filter; it worked as expected. Because I have not seen publications in the last 30 years on this unique filter topology, I decided to write an article about it for Linear Audio. The next part is based on our old hand-written notes with some additional simulations and examples. And, as a bonus, at the end of this article an alternative Linkwitz Transform filter is presented.

2 The Sallen-Key biquad filter

Figure 1 shows the general topology of the Sallen-Key biquad filter.

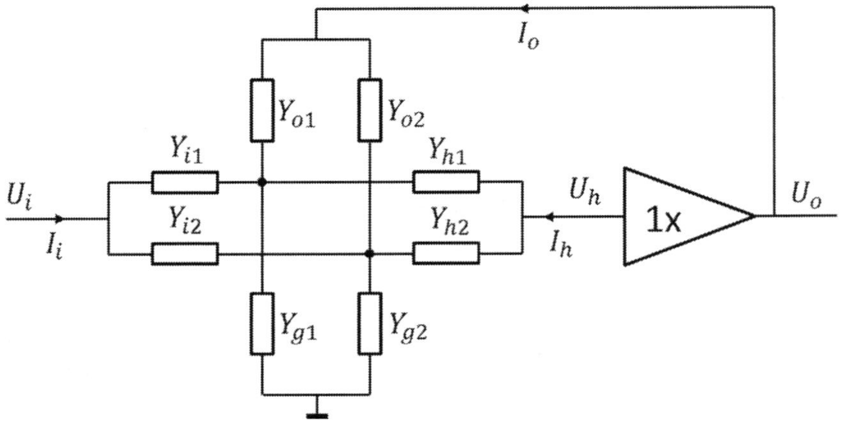

Figure 1: General topology of the Sallen-Key biquad filter

The name biquad is short for biquadratic, as a biquad filter is a type of linear filter that implements a transfer function that is the ratio of two quadratic functions. Let's try to derive the transfer function that we can use to dimension the circuit. Upfront warning: despite the relatively low number of components, the math becomes extensive very quickly. We can define the following admittance matrix:

$$\begin{bmatrix} I_o \\ I_i \\ I_h \end{bmatrix} = \begin{bmatrix} Y_{oo} & Y_{oi} & Y_{oh} \\ Y_{oi} & Y_{ii} & Y_{ih} \\ Y_{oh} & Y_{ih} & Y_{hh} \end{bmatrix} \cdot \begin{bmatrix} U_o \\ U_i \\ U_h \end{bmatrix} \tag{1}$$

Using an ideal buffer, we can set $I_h = 0$, so we get:

$$0 = Y_{oh}U_o + Y_{ih}U_i + Y_{hh}U_h \tag{2}$$

Because the buffer is unity-gain, $U_o = U_h$. So we get the transfer function:

$$\frac{U_o}{U_i} = \frac{-Y_{ih}}{Y_{oh} + Y_{hh}} \quad \text{with:} \tag{3}$$

$$Y_{oh} = \left.\frac{I_h}{U_o}\right|_{U_i=U_h=0} = \frac{-Y_{o1}Y_{h1}}{Y_{o1} + Y_{i1} + Y_{h1} + Y_{g1}} + \frac{-Y_{o2}Y_{h2}}{Y_{o2} + Y_{i2} + Y_{h2} + Y_{g2}} \tag{3a}$$

$$Y_{ih} = \left.\frac{I_h}{U_i}\right|_{U_o=U_h=0} = \frac{-Y_{i1}Y_{h1}}{Y_{o1} + Y_{i1} + Y_{h1} + Y_{g1}} + \frac{-Y_{i2}Y_{h2}}{Y_{o2} + Y_{i2} + Y_{h2} + Y_{g2}} \tag{3b}$$

$$Y_{hh} = \left.\frac{I_h}{U_h}\right|_{U_o=U_i=0} = \frac{Y_{h1}(Y_{o1} + Y_{i1} + Y_{g1})}{Y_{o1} + Y_{i1} + Y_{h1} + Y_{g1}} + \frac{Y_{h2}(Y_{o2} + Y_{i2} + Y_{g2})}{Y_{o2} + Y_{i2} + Y_{h2} + Y_{g2}} \tag{3c}$$

Substituting (3a), (3b) and (3c) in (3) yields:

$$\frac{U_o}{U_i} = \frac{\dfrac{Y_{i1}Y_{h1}}{Y_{o1} + Y_{i1} + Y_{h1} + Y_{g1}} + \dfrac{Y_{i2}Y_{h2}}{Y_{o2} + Y_{i2} + Y_{h2} + Y_{g2}}}{\dfrac{Y_{h1}(Y_{i1} + Y_{g1})}{Y_{o1} + Y_{i1} + Y_{h1} + Y_{g1}} + \dfrac{Y_{h2}(Y_{i2} + Y_{g2})}{Y_{o2} + Y_{i2} + Y_{h2} + Y_{g2}}} \tag{4}$$

For simplification of (4), we introduce a constant k, which is to be defined:

$$Y_{o2} + Y_{i2} + Y_{h2} + Y_{g2} = k(Y_{o1} + Y_{i1} + Y_{h1} + Y_{g1}) \tag{5}$$

Now we can rewrite (4) as follows:

$$\frac{U_o}{U_i} = \frac{kY_{i1}Y_{h1} + Y_{i2}Y_{h2}}{kY_{h1}(Y_{i1} + Y_{g1}) + Y_{h2}(Y_{i2} + Y_{g2})} \tag{6}$$

If we only use combinations of conductances (the inverse of a resistance) and capacitances for the admittances, then equation (6) gets a second order polynomial in $s = j\omega$ in both the numerator and the denominator, and that is what we need. The following parts offer good possibilities:

$$\begin{array}{llll} Y_{i1} = G_{ci} + sC_i & Y_{g1} = G_{cg} + sC_g & Y_{h1} = sC_h & Y_{o1} = G_o \\ Y_{i2} = G_i & Y_{g2} = G_g & Y_{h2} = G_h & Y_{o2} = sC_o \end{array}$$

This results in the filter of **figure 2**.

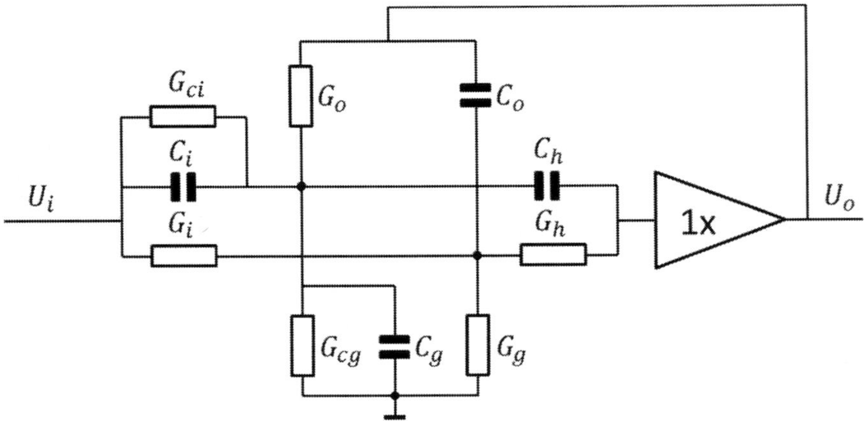

Figure 2: Basic biquad filter

Substituting these parts in (6) yields:

$$\frac{U_o}{U_i} = \frac{C_i}{C_i + C_g} \cdot \frac{s^2 + s\frac{G_{ci}}{C_i} + \frac{G_i G_h}{kC_i C_h}}{s^2 + s\frac{G_{ci} + G_{cg}}{C_i + C_g} + \frac{G_h(G_i + G_g)}{kC_h(C_i + C_g)}} \tag{7}$$

When we substitute these admittances also in (5), we get by setting identical:

$$k(G_o + G_{ci} + G_{cg}) = G_i + G_h + G_g \tag{8}$$

$$k(C_i + C_h + C_g) = C_o \tag{9}$$

The filter can be described by the following transfer function:

$$H(s) = K\frac{s^2 + sa_1 + a_0}{s^2 + sb_1 + b_0} \tag{10}$$

By comparison with (7):

$$a_1 = \frac{G_{ci}}{C_i} \tag{11}$$

$$a_0 = \frac{G_i G_h}{kC_i C_h} \tag{12}$$

$$b_1 = \frac{G_{ci} + G_{cg}}{C_i + C_g} \tag{13}$$

$$b_0 = \frac{G_h(G_i + G_g)}{kC_h(C_i + C_g)} \tag{14}$$

It appears that we should not make $C_i = 0$ or $C_h = 0$, because that would make a denominator zero; these components C_i and C_h are clearly necessary for such a biquad filter.

From (14) follows:

$$G_h = \frac{b_0 k C_h (C_i + C_g)}{G_i + G_g}$$

And from (8) and (13) follows:

$$G_i + G_g = k(G_o + G_{ci} + G_{cg}) - G_h = k[G_o + b_1(C_i + C_g)] - G_h$$

From the last two equations follows:

$$G_h^2 - G_h[kG_o + b_1 k(C_i + C_g)] + b_0 k C_h (C_i + C_g) = 0$$

We can solve this quadratic expression in G_h:

$$G_h = \frac{b_1 k(C_i + C_g) + kG_o}{2} \pm \frac{1}{2}\sqrt{k^2[b_1(C_i + C_g) + G_o]^2 - 4b_0 k C_h (C_i + C_g)} \tag{15}$$

From (11), (12), (13) and (14) follows:

$$G_{ci} = a_1 C_i \tag{16}$$

$$G_{cg} = (b_1 - a_1)C_i + b_1 C_g \tag{17}$$

$$G_i = \frac{a_0 k C_h C_i}{G_h} \tag{18}$$

$$G_g = \frac{kC_h}{G_h}[(b_0 - a_0)C_i + b_0 C_g] \tag{19}$$

$$C_o = k(C_i + C_h + C_g) \tag{20}$$

Because in (15) the square root term must be positive, we know that the minimum value for k is:

$$k_{min} = \frac{4b_0 C_h (C_i + C_g)}{[b_1(C_i + C_g) + G_o]^2} \tag{21}$$

3 The biquad filter as inverse resonance filter

We return to (10), the transfer function of the filter. What values of a_1, a_0, b_1 and b_0 do we actually need to get an inverse resonance filter? What is the desired frequency and phase response? The constant K in (10) is just a gain factor and does not change the shape of the frequency response. **Figure 3** shows the basic target transfer function for a phono cartridge inverse resonance filter.

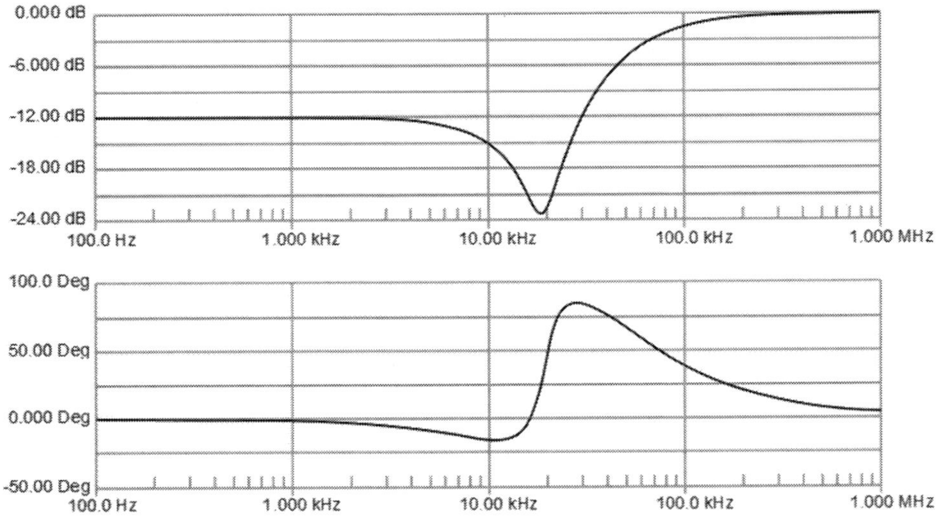

Figure 3: Target transfer function

We only have to design for the desired amplitude response. Because the network is minimum phase, the phase response is a direct result of the chosen amplitude response. At first glance it seems right to strive for exactly the inverse second order response of the mechanical resonance, which means that after the dip the response should continue to go up with 12 dB/oct. However, this would have a negative effect on the signal to noise ratio and overload margin. Actually, we are more interested in the proper amplitude and phase response around 20 kHz than in the response at much higher frequencies. Therefore we do not only design a complex conjugate pair of zeroes at the resonant frequency ω_{0z}, but also a conjugate pair of poles at a higher frequency ω_{0p} to flatten the frequency response again. Let us assume $\omega_{0p} = 2\omega_{0z}$ as a practical ratio. To get an idea about the meaning of the values a_1, a_0, b_1 and b_0, we write the numerator and denominator of (10) in their normalized form:

$$\left(j\frac{\omega}{\omega_0}\right)^2 + \left(j\frac{\omega}{\omega_0}\right)\frac{1}{Q} + 1$$

In the numerator this means that $a_0 = 1$ and $a_1 = 1/Q_z$. Because we chose $\omega_{0p} = 2\omega_{0z}$, we get $b_0 = 2^2 a_0 = 4$. I want the filter to become flat after ω_{0p} with a critically damped response, so $Q_p = 1/2$. This means that $b_1 = 2 \cdot 1/Q_p = 4$. If you want a Butterworth response instead of a critically damped response, then use $Q_p = 1/\sqrt{2}$, for a Bessel response take $Q_p = 1/\sqrt{3}$. Filling in these values in (10) gives us the basic normalized transfer function of the inverse resonance filter:

$$H(s) = K\frac{s^2 + \frac{1}{Q_z}s + 1}{s^2 + 4s + 4} \tag{22}$$

At this point it appears that there are no real objections against making $C_g = 0$, so removing C_g from the circuit. This is possible because $(b_1 - a_1) > 0$ for practical values of Q_1 and $(b_0 - a_0) > 0$, so G_{cg} and G_g will remain positive anyway.

Now we can define the admittances. There is no objection against setting $C_i = 1$ and also $C_h = 1$, and save another component by setting $G_o = 0$. Using these values in (21), we see that $k_{min} = 1$. For simplification it's fine to fix k to $k = 1$. Then, using (20), we get $C_o = 2$. Furthermore, via (16) we get $G_{ci} = 1/Q_z$ and via (17) we get $G_{cg} = 4 - 1/Q_z$. Because $k = 1$, the square root term in (15) becomes zero and we get a single value for G_h, namely $G_h = 2$. Then via (18) it follows that $G_i = 1/2$, and via (19) it follows that $G_g = 3/2$.

Now it's time to convert the conductances to resistances, using $R = 1/G$. We get the following parts list for this normalized filter:

$C_g = 0$	$C_h = 1$	$R_{ci} = Q_z$	$R_i = 2$	$R_g = 2/3$
$C_i = 1$	$C_o = 2$	$R_{cg} = Q_z/(4Q_z - 1)$	$R_h = 1/2$	$R_o = \infty$

The circuit is now quite simple and very easy to dimension, see **figure 4**.

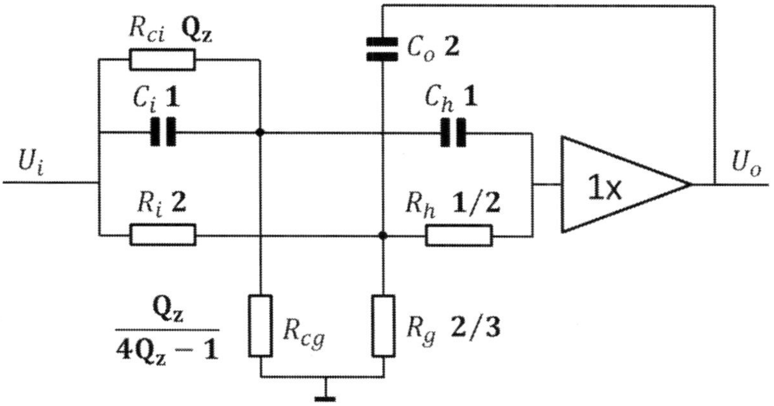

Figure 4: Normalized values of inverse resonance filter

What remains is to calculate the constant K. We can rewrite equation (7), using $s = j\omega$, $R = 1/G$ and $C_g = 0$.

$$\frac{U_o}{U_i} = \frac{(j\omega)^2 + (j\omega)\dfrac{1}{R_{ci}C_i} + \dfrac{1}{R_i R_h C_i C_h}}{(j\omega)^2 + (j\omega)\dfrac{1}{(R_{ci}//R_{cg})C_i} + \dfrac{1}{R_h(R_i//R_g)C_i C_h}}$$

$$\frac{U_o}{U_i} = \frac{(R_i//R_g)}{R_i} \cdot \frac{(j\omega)^2 R_i R_h C_i C_h + (j\omega)\frac{R_i R_h C_h}{R_{ci}} + 1}{(j\omega)^2 R_h (R_i//R_g) C_i C_h + (j\omega)\frac{R_h (R_i//R_g) C_h}{(R_{ci}//R_{cg})} + 1} \tag{23}$$

We see that $K = (R_i//R_g)/R_i = R_g/(R_i + R_g) = 1/4$. This is a 12 dB attenuation factor; not a value that we can choose ourselves, but simply the effect of selecting $\omega_{0p} = 2\omega_{0z}$ and having a second order slope of 12 dB/oct. As this filter is built around a unity-gain buffer, it cannot amplify and its maximum gain is 0 dB. Of course, the expression $(R_i//R_g)$ is short for R_i in parallel with R_g, which equals $1/(1/R_i + 1/R_g)$.

To check that we did not make any mistakes in (23), we calculate the filter parameters again from (23) and substitute the normalized parts:

$$\omega_{0z} = \sqrt{\frac{1}{R_i R_h C_i C_h}} = 1 \qquad Q_z = \frac{R_{ci} C_i}{\sqrt{R_i R_h C_i C_h}} = Q_z$$

$$\omega_{0p} = \sqrt{\frac{1}{R_h (R_i//R_g) C_i C_h}} = 2 \qquad Q_p = \frac{(R_{ci}//R_{cg}) C_i}{\sqrt{R_h (R_i//R_g) C_i C_h}} = \frac{1}{2}$$

These results are exactly as expected.

Nowadays, it is easy and quick to do a simulation of the filter on a PC and see whether the filter behaves as expected. Back then in 1981 we did not have a PC, but my friend was the proud owner of a programmable calculator with magnetic card reader, an SR52 made by Texas Instruments. So we derived the expressions for the Bode amplitude and phase diagrams and entered these in the programmable calculator to draw the Bode plots by hand:

$$|H(\omega)| = \frac{R_g}{R_i + R_g} \cdot \sqrt{\frac{(1 - \omega^2 R_i R_h C_i C_h)^2 + \left(\omega \frac{R_i R_h C_i}{R_{ci}}\right)^2}{\left(1 - \omega^2 \frac{R_i R_g}{R_i + R_g} R_h C_i C_h\right)^2 + \left(\omega \frac{R_i R_g (R_{ci} + R_{cg})}{(R_i + R_g) R_{ci} R_{cg}} R_h C_h\right)^2}} \quad \text{and}$$

$$\Phi(\omega) = \arctan \frac{\omega \frac{R_i R_h C_h}{R_{ci}}}{1 - \omega^2 R_i R_h C_i C_h} - \arctan \frac{\omega \frac{R_i R_g (R_{ci} + R_{cg})}{(R_i + R_g) R_{ci} R_{cg}} R_h C_h}{1 - \omega^2 \frac{R_i R_g}{R_i + R_g} R_h C_i C_h}$$

4 Dimensioning the inverse resonance filter

Let's take for example a filter with $f_0 = 18.5$ kHz and $Q_z = 3$. Now we can choose a convenient value for the capacitors and see whether the resistors get practical values. Suppose $C_0 = 1.5$ nF;

this results in $R_0 = 1/2\pi f_0 C_0 = 5.6$ kΩ. Multiplying by the normalized values as shown in figure 4, we get as result: $C_i = 1.5$ nF, $C_h = 1.5$ nF, $C_o = 3$ nF, $R_{ci} = 16.8$ kΩ, $R_{cg} = 1.5$ kΩ, $R_i = 11.2$kΩ, $R_h = 2.8$ kΩ and $R_g = 3.7$ kΩ. Exactly these components were used to create the diagrams of figure 3, but this time I used the simulation program CircuitMaker 2000 to get these diagrams with a lot less effort.

Notice that the dip frequency is not exactly at f_0 but comes slightly before the transition, which means that for low Q values the frequency of the minimum is somewhat lower. This is clearly visible in **figure 5**, where the transfer is shown for twelve values of Q between 0.7 and 7, all at the same resonant frequency.

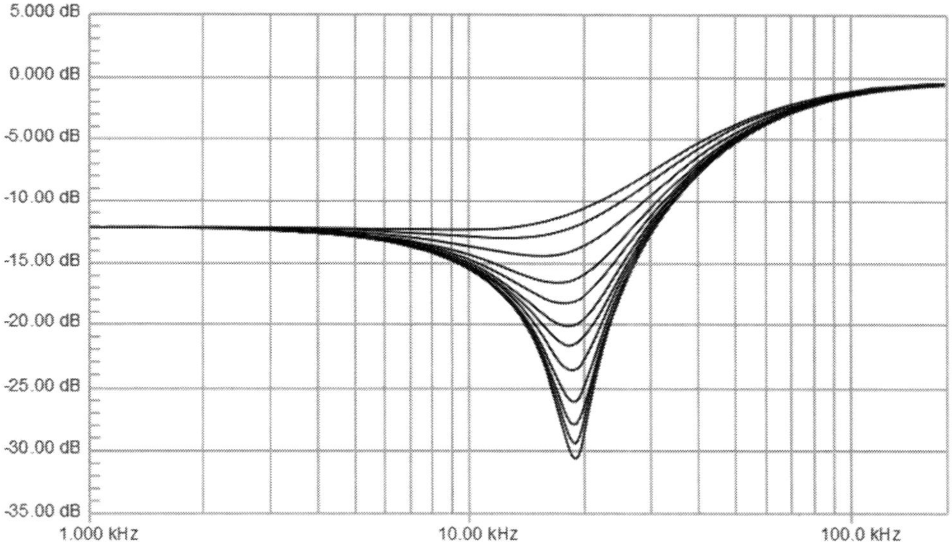

Figure 5: Various values of Q for the same resonant frequency

We also need to know the relationship between the value of Q and the depth of the dip. This relationship is complicated, because the effect of the added poles at twice the resonant frequency must be taken into account and the dip is not exactly at the resonant frequency. Instead of trying to derive an expression for this relationship, it is easier to simulate the response for a number of Q values and enter the depth of the dip in an Excel spreadsheet. The result is shown in **figure 6**. Even though the dip depth is the result of the selected Q, it makes sense to reverse the axes, so it looks like the Q is now the result of the selected dip depth. In most cases you want to create a dip of a certain depth and need to know what Q should be used for the filter. Excel provides the possibility to calculate a trend line and the shown third order polynomial matches perfectly with the results of the simulations. You can use that formula to calculate the Q for a certain dip depth.

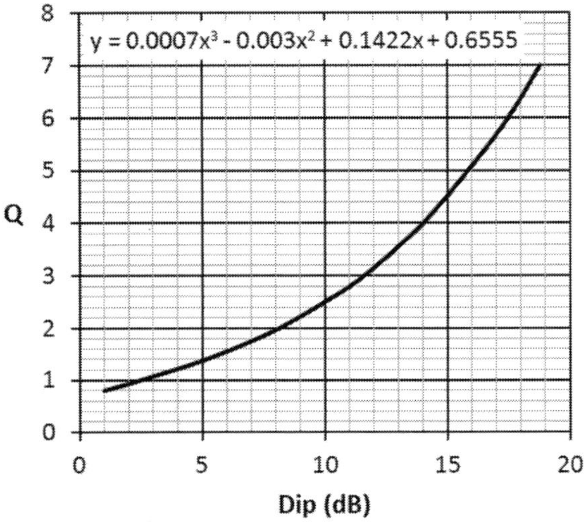

Figure 6: Quality factor as function of dip depth

5 Practical example using a real cartridge

Let's take a real phono cartridge and check its published technical data, see **photo 1**. Originally we used the Ortofon VMS20E Mk II, but this cartridge is now obsolete. Hans van Maanen used the Stanton 681 EEE cartridge as example in his articles, and the Mk III version is still for sale. This phono cartridge has a good track record (pun intended) and appears to be a good filter candidate too. For an overview of cartridges with their technical data, you can check [4].

Stanton 681 EEE MK III Technical Specifications (with special elliptical stylus)

Stylus Type:	Stereohedron elliptical 0.3 x 0.7
Frequency Response:	20 Hz to 22 kHz
Nominal Output @ 1 kHz:	2.5 mV
Channel Separation @ 1 kHz:	35 dB
Channel Balance @ 1 kHz:	within 2 dB
Tracking Force:	0.75 to 1.5 grams
DC Resistance:	1300 Ω nom
Inductance:	930 mH
Cartridge Weight:	6.3 g
Tracking Ability:	80 μ @ 3 grams
Recommended Load:	47 kΩ and 275 pF

Photo 1: Technical data of the Stanton 681 EEE Mk III cartridge

We see that this cartridge has an inductance of 930 mH, a coil DC resistance of 1300 Ω and should be loaded with 275 pF and 47 kΩ to get a 'flat' frequency response up to 22 kHz. The combination of the coil inductance and the load capacitance and resistance creates a second order low pass filter with a corner frequency of 11 kHz, being 12 dB down at 20 kHz. Only the peak of the mechanical resonance compensates for this roll-off to make the frequency response reasonably flat, except for some wiggles. We can calculate the roll-off of the LCR filter (ignoring the DC series resistance of the coil) using:

$$|H(\omega)| = 20\,log\frac{1}{\sqrt{1+\omega^2\left(\frac{L^2}{R^2}-2LC\right)+\omega^4L^2C^2}}$$

But it is easier to put these values in a simulator. In **figure 7** curve E is the electrical roll-off due to the high inductance of the coil in combination with the cable capacitance and amplifier input resistance.

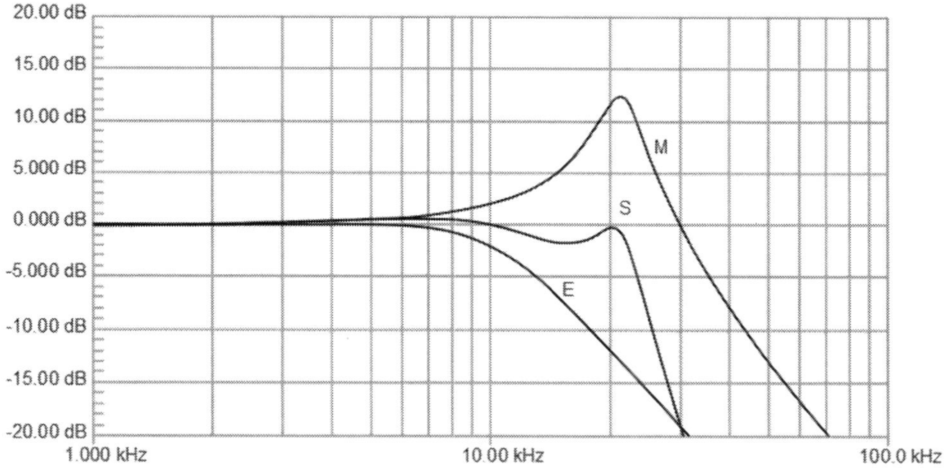

Figure 7: Simulated frequency response of Stanton 681 EEE

Curve M is the estimated mechanical resonance, based on the fact that curve S, which is the sum in dB of the curves E and M, is as flat as possible.

It is clear now why the frequency response of a phono cartridge is so sensitive to the characteristics of the interconnection cable. Using a cable with a larger capacitance will change the electrical roll-off quite dramatically, while the mechanical compensation remains the same. The better phono preamplifiers provide switchable load resistors and capacitors at the input to tune the frequency response for a specific cartridge. Some other amplifiers have difficulties with RF interference on their phono inputs and 470 pF ceramic capacitors are soldered directly across the inputs to satisfy the test agencies. It is clear that these capacitors and the cable capacitance of the turntable together will ruin the HF behavior of the cartridge. It was never designed to be loaded with such a high capacitance. This is much less a problem and much less the case for moving coil cartridges, which have a low inductance and are not very sensitive to load impedance. For moving coil cartridges the mechanical resonance is not needed for compensation and it is shifted to a higher frequency and/or is better damped.

Interesting to know is how this compensation and resulting 4th order roll-off (2nd order electrical plus 2nd order mechanical, together quite similar to a 4th order Chebyshev filter with 3 dB ripple) with associated phase shifting will affect the impulse or transient response. Eventually it is good to use a test record with square wave signals, but first we can simulate it. I used CircuitMaker 2000 (but you can also use LTspice or any other analog simulator tool) to model the mechanical low pass filter as a regular Sallen-Key low pass filter with a resonant frequency of 21.5 kHz and a peak of 12 dB, like it was estimated in figure 7. This filter is then followed by the electrical filter consisting of the series inductance (930 mH) and resistance (1300 Ω) of the coil and the recommended shunt impedance of 275 pF in parallel with 47 kΩ. A buffer stage is then added in order not to mess up the cartridge load, followed by the inverse resonance filter.

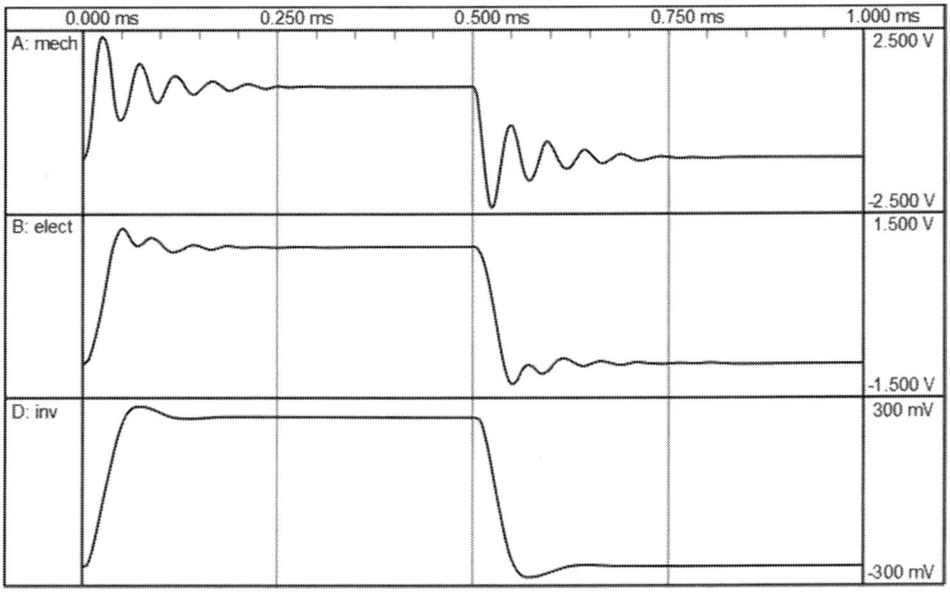

Figure 8: Cartridge square wave responses (47 kΩ load)

In **figure 8**, trace A shows the result of the mechanical resonance on the 1 kHz square wave, with severe ringing at the resonance frequency. In practice, however, most of the ringing will be filtered away by the electrical filter comprising the coil inductance with its load impedance, see trace B. Trace D is the result after the inverse resonance filter. There is still some overshoot, but this result is already a lot better. Tuning the inverse resonance filter will not remove this last bit of overshoot, as it is not a residue of the mechanical resonance. This overshoot is caused by the electrical filter that also happens to be under-damped! Fortunately this is something that we can easily solve by just increasing the load of the cartridge, so by decreasing the load resistance.

If we look at the top curve of **figure 9**, we see the electrical cartridge response for a 47 kΩ load; this is the same curve as curve E of figure 7. It is almost flat (like a Butterworth filter) up to the transition, but it is still under-damped. By changing the load resistance to a lower value, the damping of the filter improves. Another option is to decrease the load capacitance, but that is more difficult in practice. The capacitance exists in the cable, not only between turntable and amplifier but also in the arm of the turntable. Of course, you can go all the way and build a small FET source follower with phantom power supply in the head shell of the turntable arm, as a buffer stage, but this will add noise and also some weight that might affect the arm's dynamic behavior. A lower value for the load resistor is much easier and you can keep the existing cabling. All changes can be done in the amplifier. What happens if we decrease the load resistance to a lower value, say 10.7 kΩ (more on that value later)? This is shown in the bottom curve of figure 9. First the low pass filter behaves as a 1st order LR-filter. At higher frequencies the influence of the (cable) capacitance starts to play a role and the filter turns into a 2nd order LC-filter. Around 10 kHz we see the big influence of the load resistance.

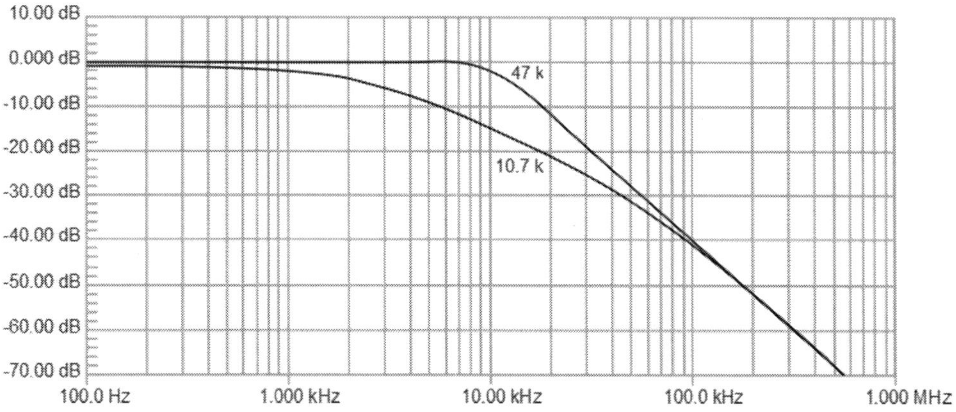

Figure 9: Electrical cartridge response for 47 kΩ and 10.7 kΩ load

A 1st order roll-off can never cause overshoot in the step response, so if we repeat the previous simulation with a 10.7 kΩ load resistor, we may expect that the overshoot will disappear. Trace A in **figure 10** shows again the ringing of the mechanical resonance on the 1 kHz square wave. Trace B shows that the electrical filter is heavily damped by the 10.7 kΩ resistor and rolls off already just above 2 kHz. An audio bandwidth of 2 kHz is not what we want and it negates all other improvements. So we have to repair the frequency response first by introducing a real zero in the chain, a 1st order high boost filter. The funny thing is that we can get this zero for free, which seems to me a fair price for a zero! All vinyl records are pressed using the RIAA pre-emphasis curve and all phono preamplifiers must have an RIAA de-emphasis network to correct the frequency response again. The time constants of this RIAA correction network are 3180 µs (pole), 318 µs (zero) and 75 µs (pole), which correspond to 50 Hz, 500 Hz and 2122 Hz. The only thing we have to do is to remove the last pole of that network, which makes that network even simpler. When the RIAA correction amplifier does not roll-off anymore after 2122 Hz, but remains flat instead, this is exactly the same as applying a high boost correction after the original RIAA correction amplifier. I did not choose the 10.7 kΩ load resistor by coincidence. Together with the coil's inductance and series resistance it introduces a pole at approximately that 2122 Hz! The addition of this high boost correction filter results in the response of trace C. The ringing from the mechanical resonance is completely back; actually it was never gone, just suppressed by the RLC filter. Because trace C is very similar to trace A, this means that our 1st order correction of the over-damped electrical RLC filter roll-off is quite good. Then the inverse resonance filter comes into place; trace D is taken at the output of that filter. It shows an almost perfect response to a square wave considering the limited frequency response up to 20 kHz. This really is a good result.

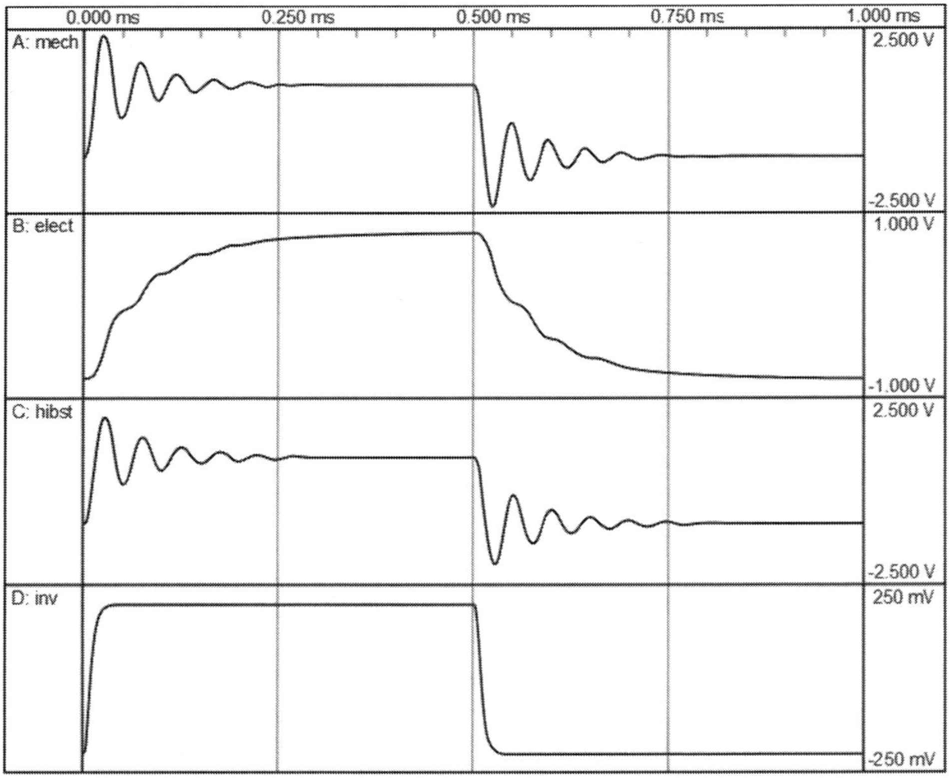

Figure 10: Cartridge square wave responses with over-damped cartridge (10.7 kΩ load)

We can calculate the required R_{load} by the following formula, based on a 1st order approximation of the network: $R_{load} = 2\pi \cdot f_c \cdot L_{coil} - R_{coil} = 2\pi \cdot 2122 \cdot 930 \cdot 10^{-3} - 1300 = 11$ kΩ.

In my simulations I needed a slightly smaller value for R_{load} to get perfectly flat lines for trace D, with the RIAA pole at 2122 Hz removed. Probably the influence of the 275 pF capacitance cannot be ignored completely, there is already a small 2nd order effect of the RLC filter. **Figure 11** shows the setup for this simulation. A Sallen-Key low pass filter with high Q to simulate the mechanical 2nd order low pass filter, followed by the electrical 2nd order low pass filter of the cartridge coil with load impedance, then the high boost correction network with time constant $\tau = R_f C_f = 75\ \mu s$ and last but not least the inverse resonance filter. The nodes A, B, C and D in the circuit are the nodes of traces A, B, C and D in figure 10 respectively. In practice the load resistor R_{load} will be part of the phono preamplifier, C_{load} will be the inevitable cable capacitance (the lower the better) and the high boost filter will just be integrated into the RIAA correction network. The inverse resonance filter is the final stage to complete this new phono preamplifier. Be aware that the inverse resonance filter has 12 dB of attenuation in the pass band that should be made up for elsewhere, either by increasing the phono preamplifier gain (take care of the overload margin) or by using an additional 12 dB gain stage after the filter.

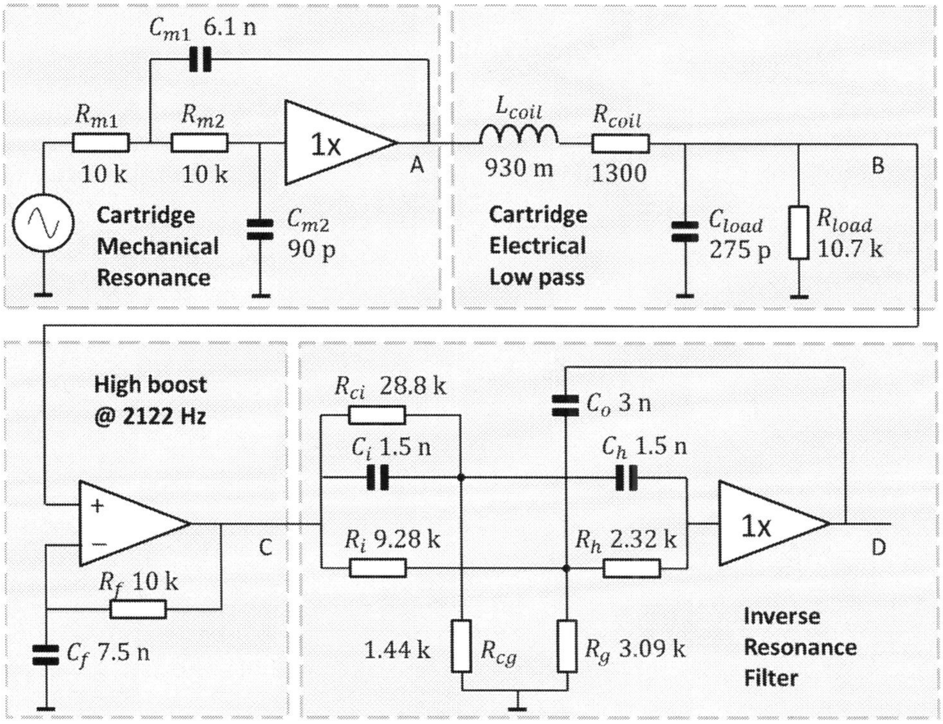

Figure 11: System setup for simulation

Of course you can load the cartridge less, for instance with 20 kΩ, and then the corner frequency will move up. Still you can use the RIAA correction amplifier to introduce an additional zero at that frequency, simply by adding a series resistor to the capacitor that causes the pole at 2122 Hz. Because of this resistor the amplifier will start to run flat again when the value of the resistor becomes bigger than the impedance of that capacitor.

The frequency responses for the nodes A, B, C and D in the system are shown in **figure 12**. Trace A shows the peaking of the mechanical filter as in figure 7, trace B shows how trace A changes due to the heavily loaded coil of the cartridge, trace C shows the result after the high boost filter, and finally trace D shows the result after applying the inverse resonance filter.

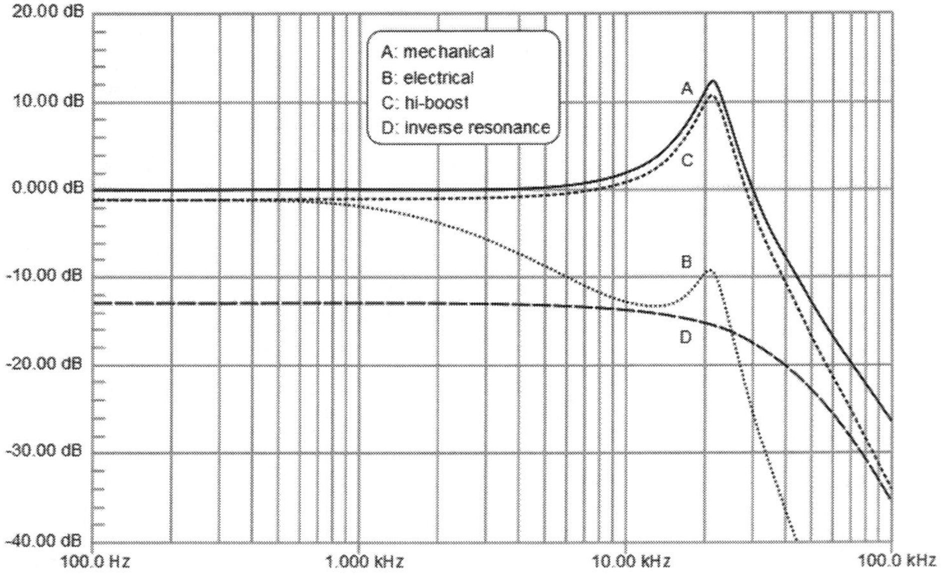

Figure 12: Frequency responses at various stages

6 A Linkwitz Transform application

Now, after having shown how to compensate for resonances at the top end of the audio spectrum and extend the useful frequency range there, it will not be such a big surprise that we can do the same at the lower end of the audio spectrum. Actually, this technique is known for years as the Linkwitz Transform [5] for active loudspeakers. Traditionally, a low frequency loudspeaker in a closed cabinet behaves as a 2nd order high pass filter with a certain resonant or corner frequency and total quality factor Q_{ts}, the latter being the combination of the electrical quality factor Q_{es} and the mechanical quality factor Q_{ms}.

The Linkwitz Transform is a biquad filter that places a pair of complex zeroes on top of the pair of poles of the high pass filter (same resonant frequency, same Q), and adds a new pair of poles at a lower frequency and typically also with a lower Q, to extend and improve the frequency response. The response of a closed box woofer can be extended with something like an octave, provided that the driver has enough power handling capacity and volume displacement. Also a too high Q_{ts}, for instance as a result of an undersized cabinet, can be improved and set to, say, 0.71 to obtain a perfect Butterworth response. This works magic for the low frequency performance.

Siegfried Linkwitz provides design rules for a single amplifier biquad filter on his website and most DIY'ers simply use this filter. However, this filter relies on an opamp with high loop gain, as shown in **figure 13**. This approach does not lend itself very well to simple discrete or tube based solutions.

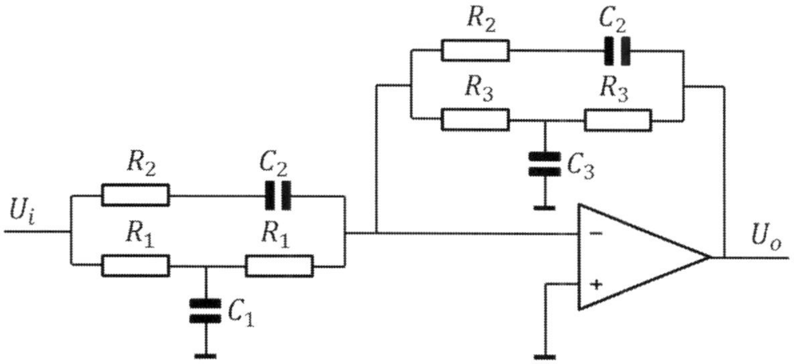

Figure 13: Traditional Linkwitz Transform circuit

I will show that the Sallen-Key biquad filter can also be used for this purpose. But as for the previous inverse resonance filter, and unlike the Linkwitz circuit, the unity gain Sallen-Key biquad filter cannot amplify, so the gain lost in making frequency response modifications has to be made up elsewhere.

While the previous inverse resonance filter has a pair of poles at a frequency above the frequency of the zeroes, now we need the opposite, and as we will see this has some impact on the filter topology. Let's start with the desired transfer function. **Table 1** shows four convenient alignments, with 6 dB or 12 dB boost for the lowest frequencies and with $Q_p = 1/2$ for critical damping or with $Q_p = 1/\sqrt{2}$ for a Butterworth alignment. The value of Q_z is to be specified (t.b.s.) and depends on the original alignment of the loudspeaker (cabinet) that needs compensation.

Filter	LT1	LT2	LT3	LT4
ω_{0z}	1	1	1	1
Q_z	t.b.s.	t.b.s.	t.b.s.	t.b.s.
ω_{0p}	$\frac{\omega_{0z}}{\sqrt{2}}$	$\frac{\omega_{0z}}{\sqrt{2}}$	$\frac{\omega_{0z}}{2}$	$\frac{\omega_{0z}}{2}$
Q_p	$\frac{1}{2}$	$\frac{1}{\sqrt{2}}$	$\frac{1}{2}$	$\frac{1}{\sqrt{2}}$
$Boost_{LF}$	6 dB	6 dB	12 dB	12 dB
a_0	1	1	1	1
a_1	$\frac{1}{Q_z}$	$\frac{1}{Q_z}$	$\frac{1}{Q_z}$	$\frac{1}{Q_z}$
b_0	$\frac{1}{2}$	$\frac{1}{2}$	$\frac{1}{4}$	$\frac{1}{4}$
b_1	$\sqrt{2}$	1	1	$\frac{1}{\sqrt{2}}$
$H(s)$	$K\frac{s^2+\frac{1}{Q_z}\cdot s+1}{s^2+\sqrt{2}\cdot s+\frac{1}{2}}$	$K\frac{s^2+\frac{1}{Q_z}\cdot s+1}{s^2+s+\frac{1}{2}}$	$K\frac{s^2+\frac{1}{Q_z}\cdot s+1}{s^2+s+\frac{1}{4}}$	$K\frac{s^2+\frac{1}{Q_z}\cdot s+1}{s^2+\frac{1}{\sqrt{2}}\cdot s+\frac{1}{4}}$

Table 1: Four Linkwitz Transform alignments

For all these filters the term $(b_0 - a_0) < 0$, so we cannot leave out C_g anymore, otherwise G_g would become negative in (19). But we can find the minimum value for C_g, which also makes $G_g = 0$, so at least we get rid of R_g. Let's determine the normalized parts for the first Linkwitz Transform filter in table 1, which is LT1. For the other filters the same procedure can be followed.

Set $C_i = 1$. Then use (19) to find that for $C_g = 1$, $G_g = 0$, irrespective of the value of k, C_h and G_h. Then we want the square root term of (15) to be zero to get a single value for G_h. Assume $k = 1$ and $G_o = 0$, then it follows that $C_h = 2$. Check (21) to find that $k_{min} = 1$, so it is allowed to set $k = 1$. Via (9) it follows that $C_o = 4$. Using (16) we get $G_{ci} = 1/Q_z$. And using (17) we get $G_{cg} = 2\sqrt{2} - 1/Q_z$. Then via (15) it follows that $G_h = \sqrt{2}$, and via (18) that $G_i = \sqrt{2}$ too. Because $R = 1/G$, we can now simply fill in the table of normalized values for this filter and in a similar way for the other versions. The result is in **table 2**.

Filter	LT1	LT2	LT3	LT4
C_i	1	1	1	1
C_g	1	1	3	3
C_h	2	1	4	2
C_o	4	3	8	6
R_{ci}	Q_z	Q_z	Q_z	Q_z
R_{cg}	$\frac{Q_z}{2Q_z\sqrt{2}-1}$	$\frac{Q_z}{2Q_z-1}$	$\frac{Q_z}{4Q_z-1}$	$\frac{Q_z}{2Q_z\sqrt{2}-1}$
R_i	$\frac{1}{\sqrt{2}}$	1	$\frac{1}{2}$	$\frac{1}{\sqrt{2}}$
R_h	$\frac{1}{\sqrt{2}}$	1	$\frac{1}{2}$	$\frac{1}{\sqrt{2}}$

Table 2: Four Linkwitz Transform normalized designs

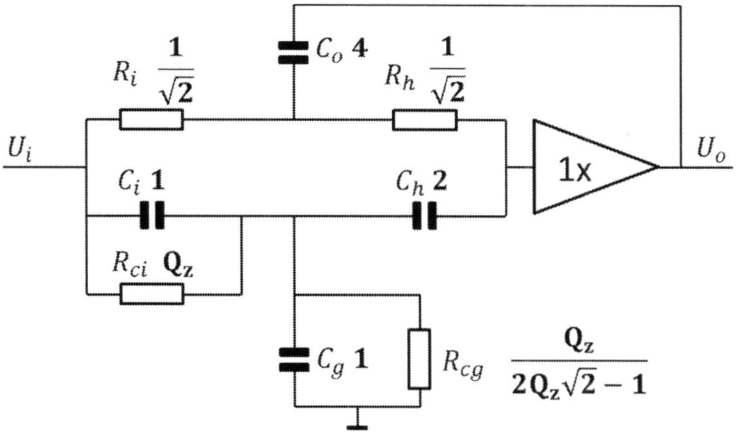

Figure 14: Linkwitz Transform example LT1

Figure 14 shows the circuit diagram with normalized component values for LT1. Use the following design procedure to calculate component values: select a convenient value for C_0, choose the required f_0 and calculate $R_0 = 1/2\pi f_0 C_0$. For instance, use $C_0 = 10$ nF and $R_0 = 200$ kΩ for $f_0 = 80$ Hz. Multiply the normalized value of each capacitor by $C_0 = 10$ nF to get the required capacitor value, and multiply the normalized value of each resistor by $R_0 = 200$ kΩ to get the required resistor value. In a practical circuit implementation add a small resistor (e.g. 100 Ω) in series with the input to avoid direct capacitive loading of the source by C_i in series with C_g. This resistor will only create a 1st order roll-off above the audio band. **Figure 15** shows the frequency responses of the four sample filters with $Q_z = 1.3$ and a resonant frequency of 80 Hz.

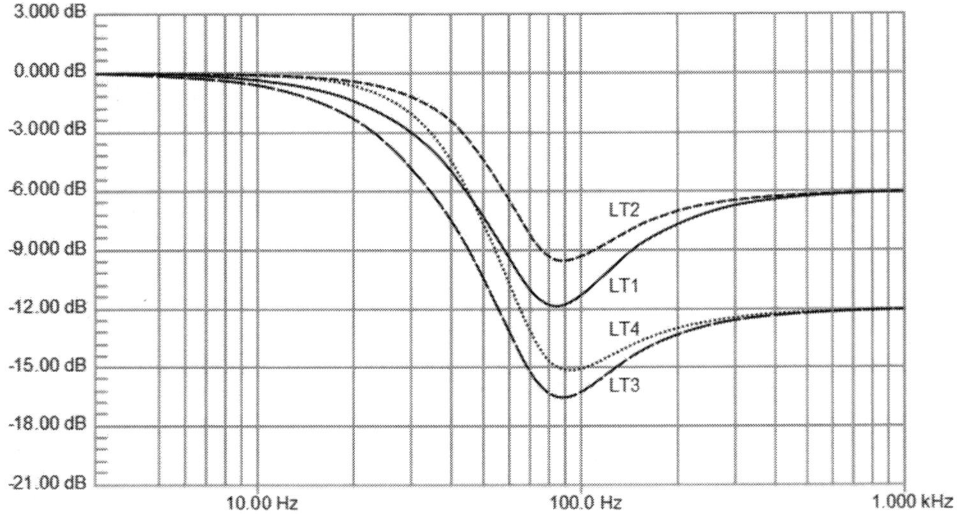

Figure 15: Linkwitz Transform filter responses

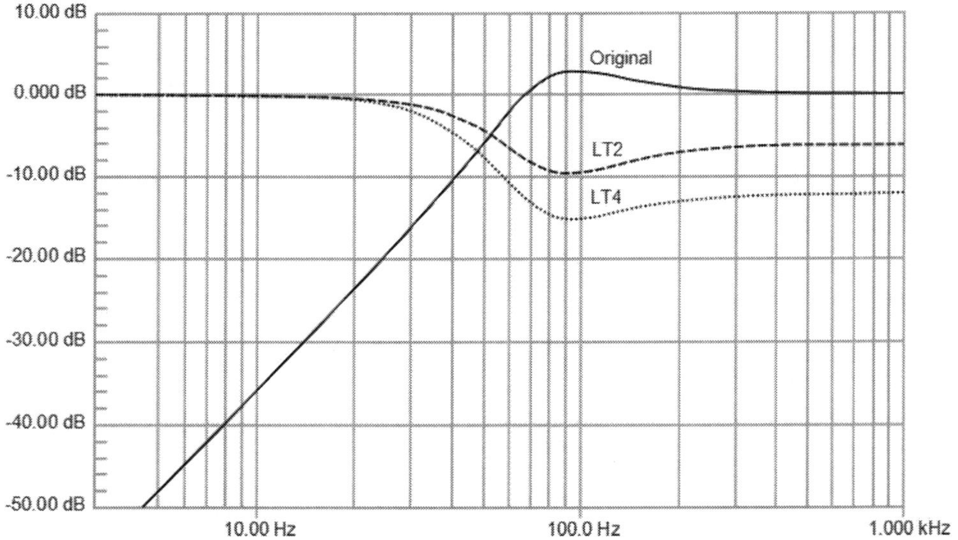

Figure 16: Under-damped closed box with LT2 and LT4 correction

Figure 16 shows the low frequency response of a closed box with $Q_{ts} = 1.3$ and a resonant frequency of 80 Hz, together with two possible corrections, LT2 and LT4. These filters change the original $Q_{ts} = 1.3$ to $Q_{ts} = 0.71$ (for Butterworth response) at a lower roll-off frequency. LT2 applies a 6 dB correction, while LT4 applies a 12 dB correction.

Figure 17 shows the total response for this loudspeaker, with and without these corrections. The response of the LT2 filter has been shifted up by 6 dB and the response of the LT4 filter has been shifted up by 12 dB to get a better comparison of the results.

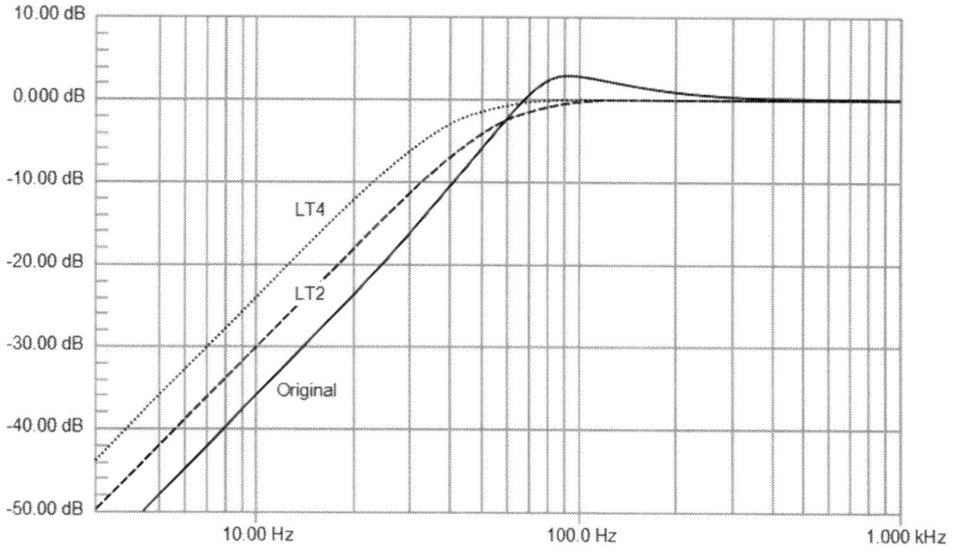

Figure 17: Low frequency response with LT4, LT2 and without filter (left to right)

This is quite an impressive result. However, keep in mind that a 6 dB correction requires 4 times the amount of power and a 12 dB correction requires even 16 times the amount of power. The amplifier should be capable of delivering this amount of power, the loudspeaker should be capable to dissipate this power and it should have enough (linear) cone excursion. Or else, the average listening level should be decreased to stay within the limits of the equipment.

7 References

[1] H.R.E. van Maanen, "Compensatie van Mechanische Resonantie bij Pick-up Elementen", Radio Elektronica 79, 15/16, pp. 25-29 and 17, pp. 35-41 (1979), in Dutch.
(download: http://www.temporalcoherence.nl/docs/MERK1979.pdf)

[2] H.R.E. van Maanen, "Signaalprocessor voor Magneto-dynamische Pick-up elementen", Radio Elektronica 80, 19, pp. 39-53 (1980), in Dutch.
(download: http://www.temporalcoherence.nl/docs/MERK1980.pdf)

[3] Prof. Dr. Ing. Rolf Unbehauen and Prof. Dr. Ing. Albert Verlag, "Netzwerksynthese in Beispielen, Band II, Passive und aktive RC-Netzwerke, RLC-Zweitore und Approximation", Aufgabe 6.17, Oldenbourg Verlag (1977), ISBN: 3-486-21571-X, in German.

[4] http://www.vinylengine.com/cartridge_database.php

[5] http://www.linkwitzlab.com/x-filters.htm#9

Passive, Discrete, DC Coupled, Open Loop IV Converters

Sigurd Ruschkowski[1]
with Erno Borbely and Karl Liander

One warm Spring day, renowned audio designer Erno Borbely asked me: „how about designing a ‚passive' IV converter for single supply, current-output DACs, and writing the design up in an article? There seem to be no such designs out there that are DC coupled". Well, I said, that sounds like great fun, but then I want it to be open loop and fully discrete and be able to handle different current output DACs like the PCM179x series and the ESS9018 chip, popularized by Twisted Pear Audio. „Sure", Erno said. And so the adventure began.
On this journey, Erno and I met a pair of IV converters which I gave the Icelandic names Regin & Fafnir and a low noise shunt regulator which I named Volsung [1].
But enough of this saga-style writing. Let´s get down to business!

Introduction

In this article we present easy-to-build solutions to the problem of converting current from differential current-output DACs into voltage. Erno Borbely and I designed a balanced output as well as a common-mode-and-even-harmonics-cancelling single ended output IV converter which both are open loop, DC coupled, and fully discrete, and which both measure and sound very good. They are also *passive*, in the sense that the initial current to voltage conversion is done passively with a resistor, followed by the active discrete circuits (therefor I loosely refer to the designs as *passive IV converters*). The designs can handle a multitude of single-supply, current-output differential DACs and they can easily be modified to suit your own design-preferences. To accompany these two open loop IV converters, an easy-to-build, fully discrete, open loop shunt regulator is shown. This shunt regulator uses a constant current source to feed the shunt, instead of the usual series resistor. Therefor, I refer to such a shunt regulator as a *super shunt* regulator.

The sources – differential, current-output, single supply DACs

High-grade differential, current output, single supply DACs for audio use have become increasingly

1 Sigurd Ruschkowski – Design and Authoring
Erno Borbely – Co-Design
Karl Liander – Thermal Design and Prototyping

popular even though the current to voltage conversion takes some more design effort than dual-supply current- and voltage-output DACs. I believe that the most popular DACs are the PCM179X-series and the ESS901X-series. I will thus focus the article´s attention to these two DAC-families.
Thinking about sending signals using currents, reminds me of my first job as SMPS designer in 1986 at RIFA Power Electronics where I started directly after having done my military service in the telecommunications signal troops. I designed a power supply for Ericsson that was to be used along oil pipelines as the oil companies needed 48VDC for their phones. To do this, the oil company sent out constant current in a cable along the pipeline. The voltage could be from 1000 VDC down to a few hundreds of volts along the pipeline. I thus had to design a SMPS that used the constant current as input and converted it into 48VDC. It was a difficult job, but together with my tutor Kjell Rolleberg, I made a design that still is in use somewhere along an oil pipeline…
Now, we will focus on the music current that comes out – in stereo mode - from the ESS9018 and PCM1792 audio DACs. If you want to run these DACs in mono mode, or for the ESS9018 use it as a multichannel DAC, it is easy to adjust the IV conversion resistor value to get the output voltage correct. More about that later on.
Both these DACs have a DC offset current. Riding on this offset current is the signal current. The PCM1792´s maximum current output is 7.8mAp-p for each phase (hot and cold), but the ESS9018 output current is is a bit more difficult to find. Using this formula (derived from the datasheet)

$$I\_signal = A_{VCC} * 0.924 / 195.31, \qquad \text{eq. (1)}$$

where the 195.31 is the output impedance of four paralleled DA converters (the ESS9018 has 16 DA single ended converters, each with an output impedance of 781.25 Ohms). With A_{VCC} = 3V3, we get

I_ signal = 15.6 mAp-p (5.5 mArms) per phase.

The DC current out from the ESS9018 – again checking the DS – is:

$$I\_dc = A_{VCC} / 2 / 195.31, \qquad \text{eq. (2)}$$

which with A_{VCC} again 3V3 gives I_dc = 8.5 mADC.
It is *these* DC currents that for some other IV converters give a DC voltage at the output making it necessary to have large DC blocking caps in series with the output signal. Our IV converters are fully DC coupled and have differential inputs so no DC blocking cap is needed.

Active or Passive Conversion?

The DAC output currents are supposed to be terminated into virtual ground. The two IV converters presented here do not present a virtual ground to the DAC output. Instead the signal current will initially be converted to voltage through a (passive) resistor.
Current output DACs want to flow into as low impedance as possible to perform their best distortion-

DAC termination	THD	Harmonics
1R1+1R1:	0.00066%	3rd only
10R+10R:	0.00073%	2nd and 3rd
15R+15R:	0.00090%	2nd and 3rd
22R+22R:	0.0012%	2nd and 3rd
205R+205R:	0.0056%	All

Table 1: Buffalo2 DAC with passive resistive loading, 24 bit 96kHz source, balanced mode, 1 kHz, 0 dBFS.

wise. Using a feedback IV conversion with an opamp one can get less than 0.1 Ohms impedance feeding the DAC current into the inverting input of the opamp. So why did we choose to use passive IV conversion? Before I answer that, let´s have a look at some THD values versus DAC termination resistor values.

Using a Buffalo2 as signal generator I created 24bits / 96 kHz sine wave WAV files playing on my SD-Trans192 version 2.1 memory card player to perform the following THD vs. termination resistor measurements (**Table 1**):

You see that THD values for 10R load resistors are not that bad, and very much the same as with an almost ideal virtual ground (the 1R1 resistor). Those low-valued resistors are very linear up into the megaherz frequencies. In my opinion, more linear than an opamp-based I/V converter when faced with very fast edges which are present on the DAC output current signal. This fact, along with the excellent distortion figures for low resistor values as shown in Table 1, made me decide to build this passive converter. OK, now let´s move on to the actual IV converter designs.

Regin – a balanced input to single ended output converter

Let's look first at a balanced to single ended, passive, open loop version I called Regin, see **figure 1**. Differential currents from the DAC (the ESS9018 in this case) are converted to a voltage by the 10R resistors R1 and R6. Choosing 10R seemed only natural looking at Table 1 and also considering that

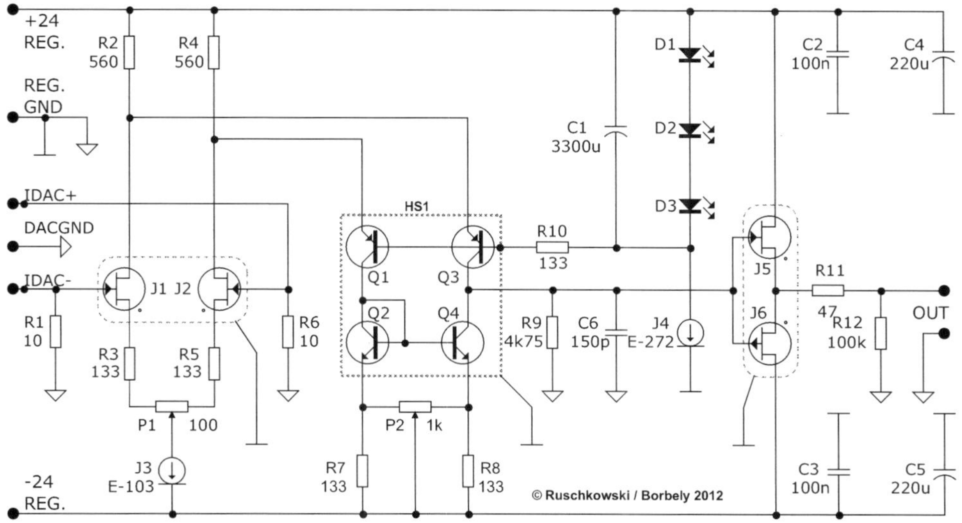

Figure 1: Balanced to single ended, passive, open loop, fully discrete and DC coupled IV Converter with conversion resistor values R1, R6 for the ESS9018 DAC in stereo mode.

the signal to noise ratio at the input voltage to the LTP will get worse with lower values; 10R is a good compromise. (The ESS9018 gives out 14.8 mAp-p into 10R instead of the above calculated 15.6 mAp-p into virtual ground because the available output voltage now is loaded by a higher total resistance. The value 195.31 in equations (1) and (2) above now has to be replaced by 195.31 + 10).

We now, thus, have 52 mVACrms at each JFET gate at 0 dBFS output from the DAC. We will also have 80 mVDC at the gates. Let´s now move on to the amplifier parts of the converters that will use these voltages as inputs.

We want to amplify these balanced AC voltages and cancel out the DC voltages, and have DC coupled, single ended / unbalanced output at 2 Vrms - into a 2k or higher load - when the DAC is fed with a 0 dBFS sine wave. I use a differential input stage loaded by a folded cascode together with a current mirror buffered by a JFET buffer. For the differential/LTP input stage (IPS), I set the JFET standing current to 5 mA. Vds for the JFETs is about 15V; a relative high voltage here is preferred because it lowers the JFET parasitic capacitances. In my opinion, but I am not sure about this (comments welcomed), there should be as little capacitance loading the DAC current outputs as possible, and then use a LP-filter later on in the IV converter to filter out any high frequency noise generated by the DAC. One also gets a bit more gate current with higher Vds, but I am not going to lose any sleep about a few nA gate current in this application. In the LTP, the 80 mVDC is a common mode voltage and will be cancelled out. But we have a DC voltage at the drains and we need to couple this stage tot he next without using coupling caps.

Enter the Folded Cascode Q1, Q3. By using a folded cascode, we can "fold" down the positive DCV at the drains. Very simple. Very effective. The folded cascode also minimises the Miller capacitance of the input stage, which will reduce capacitance modulation. So, the DAC "sees" not only low resistance using the 10R resistors, but also a low capacitance giving a total low load impedance up to very high frequencies. Another advantage of a cascode is that we get a wider bandwidth amplifier.

The bias current though the folded cascode and current mirror is 10 mA. This current, set by R2 and R4, is a good compromise between driving R9 with not too much distortion while still limiting the dissipation to acceptable levels.

Gain is set by the ratio oft he IPS drain- and source resistors and R9. The values used in Figure 1 give us 2 Vrms output voltage at 0 dBFS – the Red Book Standard voltage. You can change R9 to get lower or higher output voltage, but beware that the THD values will change due to higher output voltages as the output voltage swing increases. A first-order LP-filter is formed by C3 and R9 together with the output impedance at Q3´s collector, with a 200 kHz -3 dB point. Too high? Just increase the capacitance of C3 to get a lower F_3.

The complete converter is complemented with an output buffer - a typical Borbely JFET buffer I would say. It is simple, measures well, and sounds good; what more can one want? Maybe more drive

Freq	THD	SNR	Harmonics
100 Hz	0.0034%	89 dB	2nd and 3rd
1 kHz	0.0042%	87 dB	2nd and 3rd
10 kHz	0.013%	78 dB	2nd to 5th

Table 2: Balanced to single ended converter distortion performance, using Buffalo2 DAC outputting -1 dBFS (1.89 VACrms converter output voltage).

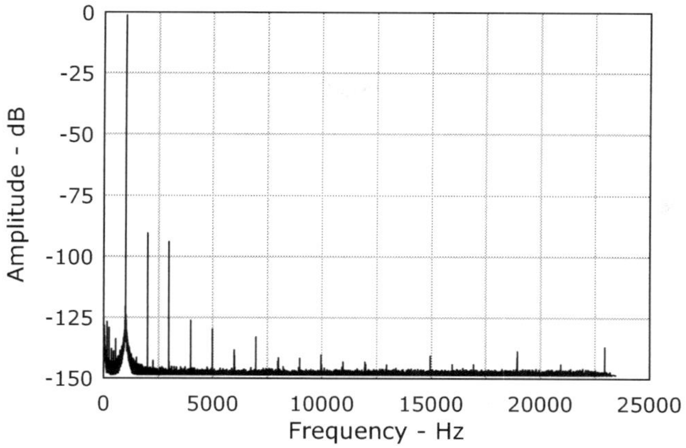

Figure 2: Balanced in to single ended out converter; 1 kHz at -1 dBFS and 1.9 VACrms Output Voltage.

capability. If so, you can use a MOSFET or BJT source/emitter follower instead of the JFET buffer.
Although we are not dealing with uV from an MC cartridge here, it is still necessary to ensure sufficiently low noise design. Most important is to use low-noise IPS JFETs - and we use Toshiba´s active production 2SK170BL's – and low value source resistors. To also get low THD from the IPS, one has to degenerate the sources – adding local feedback – and high value source resistors contradict low noise. The given values (133 ohms) are sufficiently low as to not compromise the noise performance. Total gain is about 26 dB. The bias voltage for the folded cascode, created by D1 – D3 and the constant current diode, E-272, is filtered by a low ESR electrolytic cap and a polypropylene cap.
I did distortion measurements on the complete converter, again using my Buffalo2 plus SDTrans192 memory card player playing high resolution WAV sine wave files at -1 dBFS (1.89 VACrms output voltage), **table 2**.
Not bad at all! SNR is lower at 10 kHz as I had to increase measurement BW from 20 kHz to 100 kHz to be able to also measure the higher harmonics of the 10 kHz spectrum. **Figure 2** shows a 1 kHz spectrum at -1 dBFS and 1.9 Vrms output voltage**. Figure 3** shows the decrease of THD levels with decreasing signal levels. When playing music with normal recording levels, this converter has such low THD levels, that one might one ask, why do we need *global feedback* IV converters?
All in all, the balanced input to single ended output converter Regin is a simple IV converter that has excellent low THD values and low noise. The parts list based on fig 1 is shown in **Table 3**.

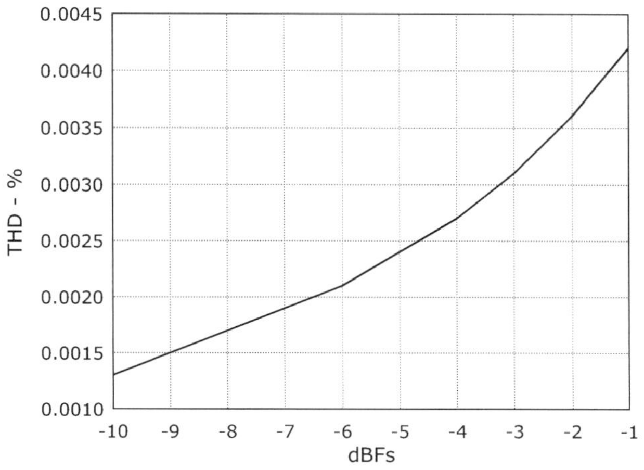

Figure 3: Balanced in to single ended out converter; THD vs. DAC output signal level

Parts	Qty.	Value	Comment
R1,R6	2	10R	Your Favorite Resistor
R12	1	100k	MetalFilm
C2,C6	2	100n/250V	PHE426
R7,R8	2	133R	MetalFilm
R5,R3,R10	3	133R	MetalFilm
C3	1	150p/100V	Polystyrene or Silver Mica
C4,C5	2	220u/35V	e.g. ELNA RJH or Pana FM
Q4,Q2	2	2SC3503	D/E/F version
J6	1	2SJ103BL	Matched to J5
J2	1	2SK170BL	Matched to J1
J1	1	2SK170BL	Matched to J2
J5	1	2SK246BL	Matched to J6
C1	1	3300u/16V	e.g. ELNA RJH or Pana FM
R11	1	47R	MetalFilm
R9	1	4k75	Your Favorite Resistor
R2,R4	2	560R	MetalFilm
J4	1	E-272	available at CPC
J3	1	E-103	Or E-452+E-562, available at CPC
Q1,Q3	2	2SA1381	D/E/F version
D2,D1,D3	3	Blue LED	3mm
P1	1	100	25T, Cermet
P2	1	1k	25T, Cermet
HS1	1	-	To fit 4 pcs TO-126 BJTs

Table 3: Parts list fort he balanced to single ended converter.

Fafnir – a balanced input to balanced output IV converter

As **Figure 4** shows, this design is balanced throughout, from input to output. The input stage is the same as for the balanced in to singled ended output version, with a 5 mA standing current in each LTP low noise 2SK170BL input JFET, with the sources in the IPS heavily degenerated for low distortion. Each hot and cold phase then has its own folded cascode using a low capacitance BJT, the 2SA1381. One current diode per phase is used to set the standing current, 10 mA, in the folded cascode. The biasing scheme creates a low noise reference voltage for the folded cascode providing adjustment for zero DC offset (P1, P2).

Gain is set by R4/R16, and there is a LP filter before and after the buffer, which is again the typical Borbely complementary JFET buffer.

Table 4 shows my THD and THD+N measurements using the above mentioned high resolution WAV files at -6 dBFS (2 VACrms output voltage).

The 100 Hz and 1 kHz results are about twice that of single ended output version, but at 10 kHz, the balanced out version is better. Comparing the spectrum from Figure 2 to **Figure 5** spectrum we see that 2nd harmonics are better suppressed in the measurement amplifier than what the current mirror in the balanced to single ended converter is capable of doing. This is to be expected as the current mirror is not as linear as the measurement amplifier. The parts list, based on Fig 4, is shown in **Table 5**.

Volsung - a discrete super-shunt regulator

Ever since I read Walt Jung´s excellent article "Don´t shun the shunt regulator" [2] from 1969, I have been a dedicated shunt-follower. However, I prefer to use a CCS to bias the shunt instead of Walt´s resistor, making it a *super-shunt*. To me, a super shunt regulator has a kind of elegance to it and re-

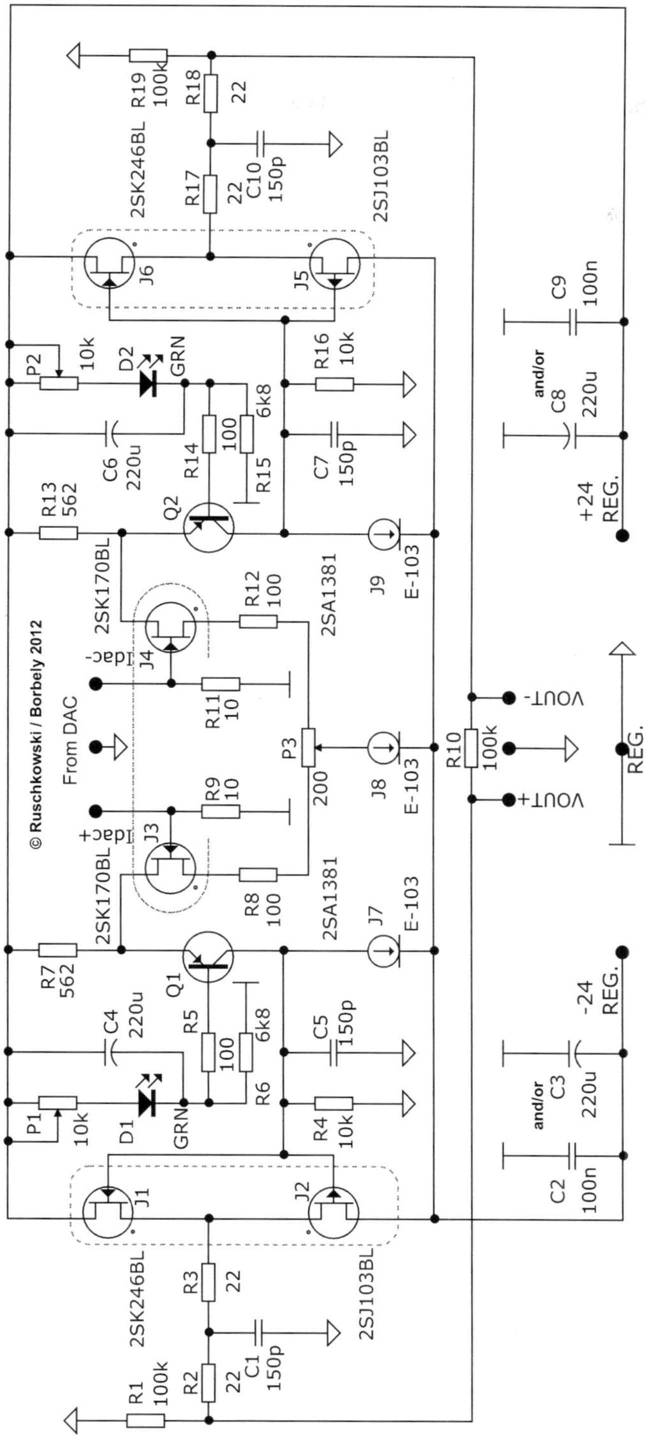

Figure 4: Balanced-in to to balanced-out, passive, open loop, fully discrete and DC coupled IV Converter with R9, R11 resistor values for the ESS9018 DAC configured for stereo mode.

Freq	THD	Harmonics	THD+N
100 Hz	0.00750%	2nd and 3rd	0.00778
1 kHz	0.00765%	2nd and 3rd	0.00776
10 kHz	0.00885%	2nd and 3rd	0.00888

Table 4: Distortion performance, using a Buffalo2 DAC at -6 dBFS (2 VACrms output voltage)

minds me of a class A amplifier with its often excellent sound qualities.
Andreas Schubert [3], "Salas" from Greece [4] and audio author and former NASA astronaut Norm Thagard [5] have all made similar super shunt regulators with excellent results.
The Volsung, see **Figure 6**, is a low noise, closed loop, fully discrete, positive and negative shunt regulator, which can be made to have a wide range of output voltages by only changing the number of LEDs in the LED-string. R1 sets the constant current supplying both the load and the shunt section, and can be adjusted for different load current requirements. In this article, Volsung is optimised for the Regin and Fafnir IV converters described above.

The cascoded CCS sources about 150mA current which is then sunk by the load and the shunt. About 100mA is sunk by the shunt MOSFET; at this relative high idle current, the shunt MOSFET has higher gm which means better voltage regulation and lower output impedance. Nominally 50mA is then available for the Regin or Fafnir which is what they need. A few blue LEDs and a pot set the output voltage. This reference voltage is well-filtered by the low pass filter R5/R6 and C7/C8 plus C5/C6, and is then fed to the control amp (Q5, Q6). The current diodes (E-452) lighty load Q5, Q6 for high gain (about 80 dB), and thus good voltage regulation. R9 and R10 (1.0 Ohms) are only there to allow a check of the shunt current and can be shorted out in actual use.
Stability is assured by two compensation networks: C9/C10 together with the Zoebel network R11/R12 plus C11/C12. The gate resistors also add to stability. Depending on how the load looks like, you can change the Zoebel network and/or add a load capacitor to further improve stability.
A shunt regulator´s main drawback is its power dissipation. I have used one large heat sink for Q1+Q3 (Q2+Q4) and another heat sink for the MOSFET shunt element Q7 (Q8). The Fischer SK104 50 can dissipate the power without getting too hot.
Many shunt regulators show a large 'reservoir' capacitor at the output. The reasoning often is that any transient loads would be sourced by that cap rather than from the shunt. I don't believe this too be the case. At full level on a 10k load, the supply (and shunt) current varies only a mA or so on a DC current of about 40mA. So any transients would be very low. Also, large capacitors stop behaving like capacitors at higher frequencies any-

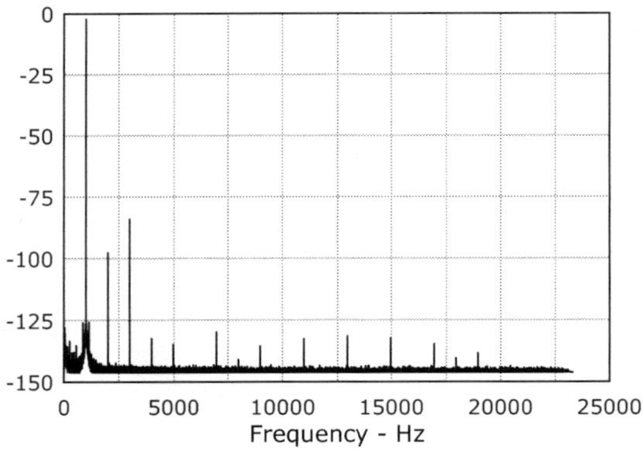

Figure 5: Balanced in to balanced out converter; 1 kHz Spectrum at -6 dBFS and 2 VACrms Output Voltage.

Ref.	Qty.	Value	Comments
R11,R9	2	10R	Your Favorite Resistor
R12,R8,R14,R5	4	100	MetalFilm
R10,R1,R19	3	100k	MetalFilm
C2,C9	2	100n	PHE42663VDC
R4,R16	2	10k	Your Favorite Resistor
C5,C1,C7,C10	4	150p	Polystyrene 100V
R3,R2,R17,R18	4	22R	MetalFilm
C6,C4	2	220u/16V	e.g. ELNA RJH or Pana FM
C8,C3	2	220u/35V	e.g. ELNA RJH or Pana FM
J5,J2	2	2SJ103BL	Matched to J6 and J1
J4	1	2SK170BL	Matched to J3
J3	1	2SK170BL	Matched to J4
J6,J1	2	2SK246BL	Matched to J5 and J2
R13,R7	2	562R	MetalFilm
R6,R15	2	6k8	MetalFilm
D1	1	LED	green, 3mm
D2	1	LED	green, 3mm
J9,J8,J7	3	E-103	or E-452+E-562
Q1,Q2	2	2SA1381 D/E/F	Sanyo, Fairchild
P3	1	200R	25T, Cermet
P1,P2	2	10k	25T, Cermet

Table 5: Balanced in to balanced out IV Converter parts list

way, so the output impedance of a well-compensated shunt may actually be lower than the cap at higher frequencies. But there's another aspect not often appreciated. A large cap in parallel to the shunt does decrease the shunt's open loop gain and thus ultimately its (ideally zero) closed loop output impedance. **Figure 7** shows the effect on a simulated shunt. The loop gain was measured with the Tian method [6, 7]. You see that a large output cap (100uF) decreases loop gain at 20kHz by 20dB wich means a ten times larger Zout! Not what we need.

However, we cannot delete the cap altogether as that would lead to instability. The zoebel consisting of R11, C11 and R12, C12 neatly solves this: the loop gain remains high till up to 20kHz, while it rolls off fast above that to assure stability. The gain and phase margin with the zoebel are 56 dB and 53 deg respectively, quite respectable. Finally, **Table 6** lists the parts list for the complete regulator.

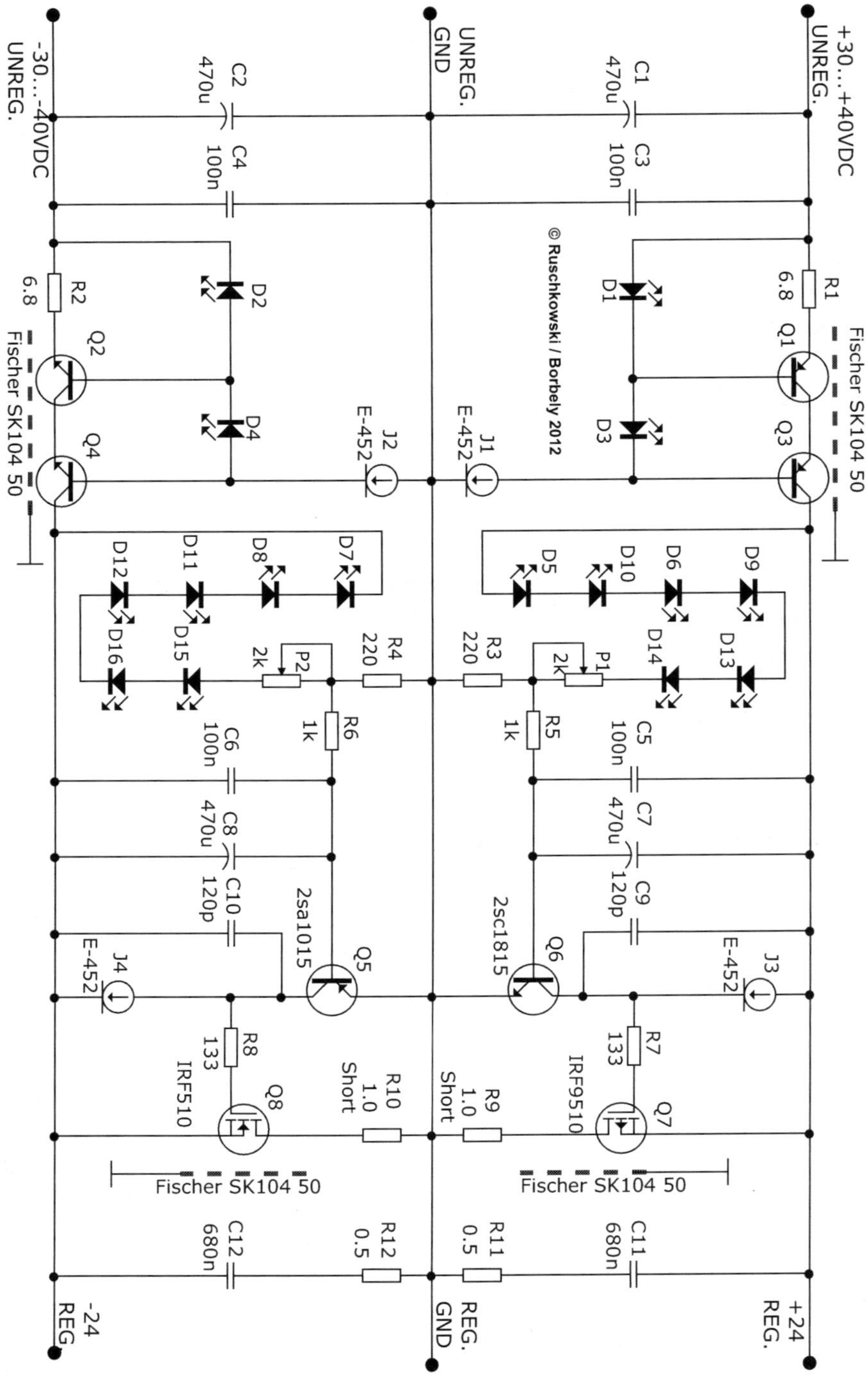

Figure 6: The Volsung - a Low Noise, Closed Loop, Fully Discrete, Super Shunt Regulator.

Applications with the current-output PCM179X family

The PCM1792´s maximum current output is 7.8 mAp-p for each phase riding on a 4 mADC current. Thus, it is half that of the ESS9018´s output current when it is configured for stereo mode. This makes it easy to make the Regin and Fafnir work for the PCM179X series. We can increase the IV resistor to nominally 20 Ohms (a standard 22 Ohms will be fine). (*Or, you can leave the 10 ohms in and increase the gain by 6 dB – ed*). If you know the output current of some other differential, single-supply DAC, you just change the IV resistor value to fit that current. With the given gain setting, you should aim for about 50mV signal across the IV resistor at 0 dBFS

Thermal Considerations and DC Offset Trimming

The advantage of having a fully DC coupled – that is no DC servo nor DC coupling cap – has its price, and that is why we need a thermally well-balanced design (and good patience with trimming duties). With that I mean, that transistors have to be thermally coupled – where appropriate – and we need a thermal equilibrium for the complete IV converter with minimal air movement. We also need to trim DC offset with the pot(s) at this thermal equilibrium. The better we are at doing this, the less DC offset and offset drift over time we will get. Karl Liander did a great job finding out a way to get thermally well balanced designs with low DC offset with low drift.

Karl put all BJTs on a small heat sink and used heat shrink tubing to put the input and buffer JFETs in close thermal contact with each other (**Photo 1**). He also mounted the input JFETs under this BJT

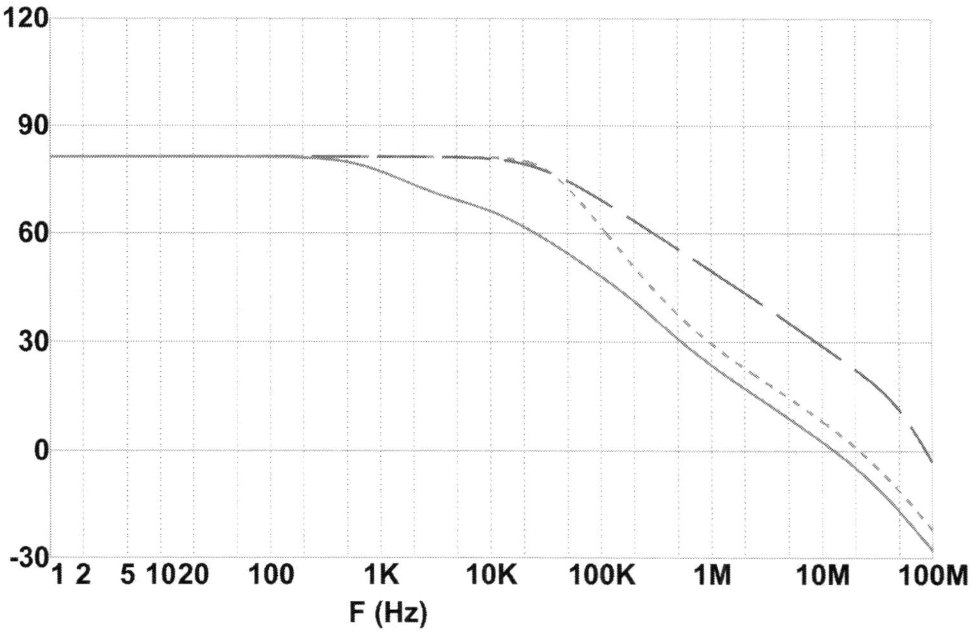

Figure 7: Simulated shunt regulator loop gain. Large dashed curve: open loop, unstable; small dash: R11, C11 zoebel, high OL gain and stable; solid: large output cap, stable but lower OL gain at high audio frequencies.

Ref	Qty.	Value	Comments
R11,R12	2	0.5R	3W MetalFilm
R10,R9	2	1R	MetalFilm
C4,C3,C5,C6	4	100n	PHE426/250V
C9,C10	2	120p	SilverMica 100V
R8,R7	2	133	MetalFilm
R5,R6	2	1k	MetalFilm
R4,R3	2	220	MetalFilm
Q5	1	2SA1015Y	Toshiba etc.
Q6	1	2SC1815Y	Toshiba etc.
C2,C1	2	470u	63V
C8,C7	2	470u	10V
R2,R1	2	6.8	3W MetalFilm
C11,C12	2	680n	PHE426/250V
Q8	1	IRF510	
Q7	1	IRF9510	
J1,J2	2	E-452	Buy at CPC
J4,J3	2	E-452	Buy at CPC
D3-D16	14	Blue LED	3mm
D1,D2	2	Green LED	3mm
Q4,Q2	2	MJE15034	OnSemi
Q3,Q1	2	MJE15035	OnSemi
P1,P2	2	2k	10 Turn, Cermet
HS	4	SK104 50 or similar	Heatsink

Table 6 Complete pos and neg super-shunt parts list

heat sink. The complete IV converter was then put in an air tight box with a small hole for access to the trim pot(s).

After waiting a few hours to get the whole thing into thermal equilibrium at the place it is to be used (in the equipment rack for example), I then trimmed the current mirror potentiometer for lowest DC offset (The source potentiometer should be trimmed first, see chapter on Setup). I then waited a while and re-checked DC offset and then re-trimmed DC offset until I was satisfied with the result.

For the balanced in to balanced out version, a similar approach was used; even with individual trim potentiometers for each phase, a common BJT heatsink is recommended to assure thermal tracking. After having trimmed the source resistor potentiometer, you can start trimming the two DC offset trim potentiometers. Again, this version was mounted in an air tight box with trimmer access holes. Wait a few hours and then start trimming. DC offset for the two phases are independent of each other, so you can trim them separately. After having trimmed DC offset, wait an hour or so, and then re-trim. Iterate until satisfied.

I did some measurements on the DC offset and drift. For the single ended output version, I used my oscilloscope set at 1000 seconds time/div setting, and noted that DC offset remained within +-5 mV over time.

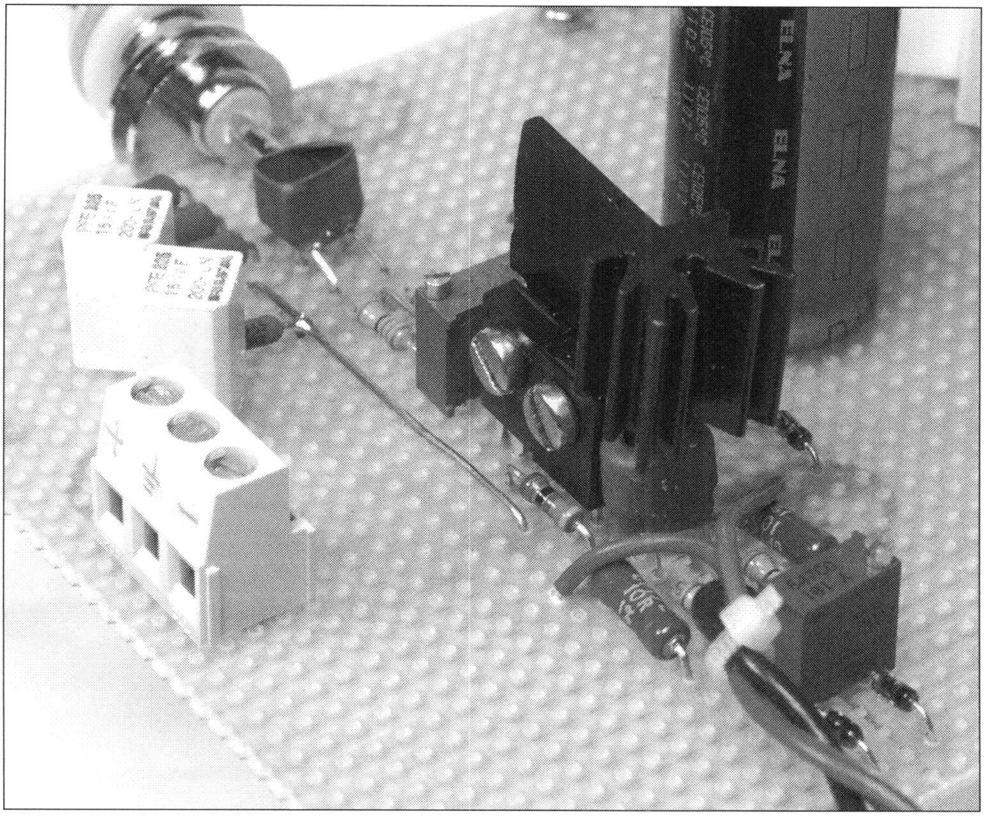

Photo 1: Suggested thermal mounting of the TO92 and TO126 transistors to assure thermal tracking.

For the balanced out converter, I noted +-20mV DC offset drift over time. The good news is that DC offset drifted in the same direction for both phases, making it common-mode which would be rejected in the balanced receiver.

Implementation and setup

The input stages of both converters use matched 2SK170BL JFETs. Matching Idss within 10% at 10V Vds is fine. The trim potentiometers will do the rest. The buffer JFETs 2Sk246BL / 2SJ103BL also have to be matched but again 10% Idss matching at 10V is fine.
The JFETs are to be thermally connected together as shown in Photo 1. Use low temperature heat shrink, and a low temperature setting on your heating gun as not to damage the JFETs.
The converters and shunt are simple enough that you can assemble them on veroboard. Of course, you can always lay out your own PCB. P2P will work, and experienced users can use this method. Our prototyping has shown that the current mirror in the balanced to single ended version is the hardest part to get right. It is very tight around the CM as all four transistors are on the same heat sink. Be

extra carefull when assembling the CM.
The most important thing about the super shunt is that large heat sinks are necessary. Except for that, just assemble it on Vero board following the schematic for positioning the parts.
After assembly, before powering up, turn all trim pots to their approximate mid position. Before you do anything else, have a break, have a glass of wine or have a beer – you did a great assembly job!
The super shunt is easy to setup. Connect +-30 VDC. After a relatively long delay, the blue LEDs will light up. Trim the output voltage to +-24VDC with the two trimpots. Measure that you have about 150 mA through the source resistors of the MOSFETs. Connect 470R power resistors across each output. If everything looks OK, you can connect an oscilloscope (if you have one) to the outputs to verify that there are no oscillations and that the output is stable.
To setup the balanced to single ended converter, connect +-24VDC from a current-limited laboratory power supply. Current draw should be about +-50mA. The three blue LEDs will shortly light up, and you can now start a rough trimming of P1. This potentiometer sets the balance of the input stage. If you have a spectrum analyser, trim P1 until you get lowest 2nd harmonics. If not, trim P1 so that the voltages across R2 and R4 are equal as best as you can. Now it is time to trim DC offset. This is a bit tedious, so make sure you are in a relaxed mood when you do this! Put the unit in an air tight box with trimmer access holes for P2. Wait a few hours so that you have thermal equilibrium. Then trim P2 for zero DC offset. Wait a while, and then re-trim. Iterate until satisfied.
The balanced in to balanced out version setup is a bit differently. Connect +-24VDC and watch the green LEDs light up. Use P1 and P2 to set DC offset to about zero. Put in an air tight box with trimmer access and again wait a few hours for thermal equilibrium. Connect your DAC to the inputs and play a 1k Hz sine wave. Trim P3 until you have the same signal level at each output phase. Then trim P1/P2 for zero DC offset at each output phase. Iterate until satisfied.
Hooking the various parts up is pretty straightforward but you might benefit from some tips: Use twisted wire to connect your DAC to the converters with as shorts leads as possible. Similarly when connecting the shunts to the converters. Use aluminium air tight boxes for the converters. The super shunt needs a well-ventilated box as the heat sinks get rather hot.

Acknowledgements

Many thanks go to Erno Borbely for pushing me to get the designs and article finished in time. Erno, you have a nice way of getting the job done without doing it yourself! And of course for giving me the privilege to cooperate with an audio legend. You may think that Erno´s just sitting around living a retired person´s comfortable life enjoying the beautiful scenery in mountain-filled Bavaria having cups of coffee with his wife all day long, but heck no! He is still active in several audio projects!
Appreciation also goes to Karl Liander, who with great enthusiasm, and with better and better quality with each iteration, kept building all those prototypes on Vero board, and for spending many hours to get the sonic imprint of the designs.
I also thank my family, Eva and Johanna and Rikard, for putting up with me when working on this article during all available spare time during the hectic November and December months.

References

[1] http://www.timelessmyths.com/norse/volsunga.html

[2] Walt Jung "Don´t shun the shunt regulator"; Electronic Design 14, July 5, 1969

[3] Andreas Schubert "Parallel regulators for audio" (The Audio Amateur, 4/90)

[4] Salas – "The simplistic Salas low voltage shunt regulator" http://www.diyaudio.com/forums/power-supplies/143693-simplistic-salas-low-voltage-shunt-regulator.html

[5] Private communication Erno Borbely – Norman Thagard

6] Michael Tian, V. Visvanathan, Jeffrey Hantgan, and Kenneth Kundert, "Striving for Small-Signal Stability", Circuits & Devices, January 2001.

[7] "Plotting Loop Gain Using the Tian Method", Spectrum Software, http://www.spectrum-soft.com/news/spring2011/loopgain.shtm

Listening Impressions and Comparison with another IV converter

Karl and I compared the Regin single-ended output version against the Borbely Audio 325 line amplifier configured as passive IV converter. The 325 is a global negative feedback, complementary, all-FET amplifier with a MOSFET OPS.

Regin was stuffed with parts I had at home. 10R bulk metal foil resistors as IV resistors and standard metal film resistors and PRP resistors were used. Only 100nF metallised propylene capacitors were used for decoupling, and PS RIFA caps were used as LP filter caps.

The 325-IV had 22R bulk metal foil resistors as IV resistors, and the other resistors were Caddock´s MK132 type. Capacitors used were radial polystyrene caps. No LP filter at the output was used. It was powered by the 254 linear regulators also by Borbely Audio.

Karl found that the Regin had an open and detailed sound, and it sounded airy. Bass was firm and deep. Sibilants were well-behaved, and the sound stage had good width and height. Dynamics and transients were excellent. On the negative side, Karl found that there was a slight lack of spatial separation of instruments and a shallower soundstage. Sound images presented themselves primarily in front of and in between the speakers. Overall, he found that the Regin was better at depicting what and how than who and where.

Myself, I found that between the Regin and the 325-IV it was a draw for an hour or so. But after even more listening, the individual sound of these two converters revealed themselves. In summary, they both sound darn good!

The 325-IV had a deeper the sound stage; Regin a tad more flat sound stage depth wise. Both had good width and height. The Regin had a punchier and heavy more sounding bass whilst the 325-IV had better articulation in the bass. While the 325-IV played hihats and tambourines with more details, Regin had more energy in the treble. I also found that Regin really opened up the sound stage (except depth wise).

A Mechatronic approach to Active Subwoofer design

Robert Munnig Schmidt

1 Introduction

Music and its reproduction have a long history in inspiring people to try to understand the factors that influence the related acoustics and perceived quality level. The theory of sound propagation by means of longitudinal pressure-velocity waves has been thoroughly described by John William Strutt, 3rd Baron Rayleigh, (aka. Lord Rayleigh, 1842 – 1919); he wrote two books on the subject: "The Theory of Sound" Vol 1 and 2. [1] These books have become quite famous and are considered as a foundation of the basic theory on vibrations and mechanical dynamics that determine the behaviour of the most advanced high-tech equipment of today.
Mechatronics is the field of engineering that deals with active controlled motion systems, consisting of a movable mechanical system, driven by one or more actuators and controlled by a feedback/feedforward controller that bases its control action on the measurement of the mechanical movement by a suitable sensor. Mechatronics in the Netherlands was first introduced at the Philips Research Laboratories in the track-following mechanism for optical recording as used in the CD player. The applied methodology became one of the success factors for the high-tech industry. Active motion feedback was also applied many years ago in the Philips **Motional feedback** loudspeakers, using integrated accelerometers to measure the movement of the loudspeaker cone. The target was to make the bass loudspeaker element act like an ideal stiff piston with a low frequency bandwidth that was less restricted by the cabinet size. Unfortunately the feedback control was only effective at low frequencies due to the dynamic break-up resonance frequencies of the membrane. The latest research to accommodate the resulting limited loop-gain of the controller was not brought to market because sales were not sufficiently successful for different reasons. The most unexpected reason was the lack of distortion. Low-frequency distortion is often appreciated by listeners because it adds "music" to otherwise inaudible sound. At low level listening a 20 Hz sine-wave is not audible but its harmonics will give the perception of a strong low bass. Another reason for the lack of success is that many (most?) audio amateurs often prefer to be free to combine any audio device while in reality the best equipment can only be designed in balance with the other parts of the audio chain. Especially the interaction between loudspeakers and power-amplifiers is important as will be shown in more

detail in this paper. One reason for the apparent need of "separates" might be that the available examples of such optimally tuned audio chains are in most cases marketed by high-margin, low-cost brands with a corresponding reduced overall quality level of the sound reproduction.
In spite of the lack of market success of these early motional feedback loudspeakers, active motion feedback is still the best way to reproduce low frequencies from a small cabinet while at the same time reducing the large levels of distortion at low frequencies that are caused by the large membrane excursions of the loudspeakers and the compression effects of the small cabinet enclosure.
As a tribute to the important role that audio reproduction has played in the early developments in the high-tech industry, this paper is the result from an instructive case study at Delft University of Technology, The Netherlands, in introductory classes on mechatronics as a clear illustration of the dynamic interaction between amplifiers and electromechanically driven mechanics. It discussed another old and well known method to achieve motional feedback without a separate sensor but instead using the interaction of the amplifier with the moving-coil actuator [2].

2 Some theoretical background

Most readers will be well involved in this matter, so only a few relevant properties of low-frequency sound reproduction are summarized here. As mentioned, sound is a physical phenomenon based on the transfer of acoustic energy by a wave phenomenon. The acoustic energy wave propagates by continuous exchange between potential energy (pressure) and kinetic energy (velocity). In air the propagation velocity of sound (c_0) equals approximately 340 m/s @ 20°C.
Sound intensity levels are primarily given in decibels, a logarithmic scale because human perception of sound intensity is logarithmic. This implies that a certain relative change of sound level is always perceived the same, independent of the actual sound level before the change. Sound levels are expressed either in intensity (power) or pressure. These aspects can be used both as they are directly interrelated. While the ^{10}logarithm of the ratio between two levels of power is expressed in "Bell" the relation expressed in decibels is:

$$X = 10\log\frac{P_1}{P_2} \text{ [dB]} \tag{2.1}$$

In electricity the power in a circuit is directly determined by Ohm's law ($V=IR$) :

$$P = VI = I^2R = \frac{V^2}{R} \text{ [W]} \tag{2.2}$$

For that reason the relation in decibel between two levels of a voltage or current signal is:

$$X = 20\log\frac{I_1}{I_2} \text{ or } X = 20\log\frac{V_1}{V_2} \text{ [dB]} \tag{2.3}$$

This definition implies that in expressing the relation between different levels of power, this relation is a factor 10 times the logarithmic ratio; between different levels of signal amplitude it is a factor of 20.

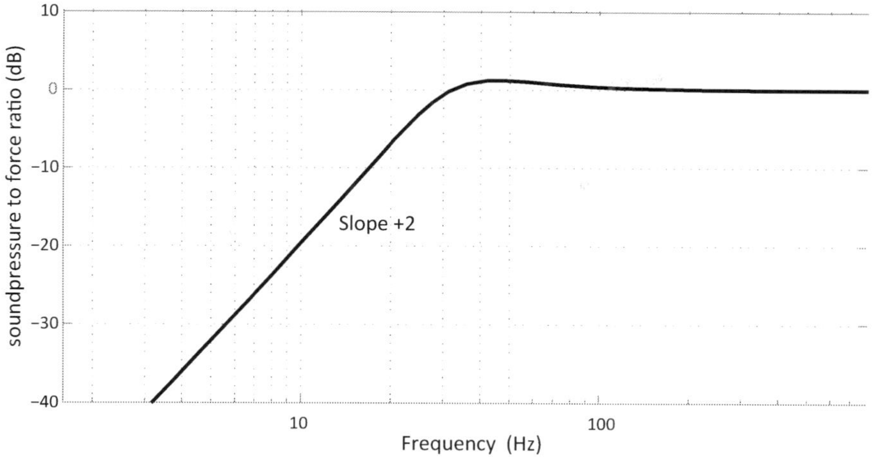

Figure 1: Magnitude Bode plot of a second-order high-pass filter where the attenuation below 40 Hz follows a positive slope of 40 dB per decade (+2).

The frequency response of a loudspeaker is generally given as a magnitude Bode-plot on a double logarithmic scale. The horizontal frequency scale covers the audio spectrum while the vertical magnitude scale is noted in decibel. Due to the above definition, the Bode-plot of the relation between the input and the output of a dynamic system like a loudspeaker is the same for either signal magnitude value or the power value.

A typical example of a magnitude Bode-plot is shown in **Figure 1**. This is the frequency response of a second-order high-pass filter that will be shown to be a good approximation of the behaviour of an electrodynamic loudspeaker in a closed box for low frequencies.

The magnitude Bode-plot is often used in acoustics while the corresponding phase-plot is not shown in most cases. The phase-plot is important in engineering when determining stability in feedback control systems but also in acoustics the phase is important when combining different loudspeakers to cover different frequency bands. Because these phase-data are hardly ever supplied by manufacturers of loudspeaker units they need to be determined by measurement or indirectly derived from the magnitude plot where a +/-1, 2 ... slope represents a +/-90, 180 ... degree phase. This relation is only valid for so called minimum-phase systems which is the case for a passive system like a loudspeaker element.

Based on the well-known Fletcher-Munson curves of equal loudness that represent the average sensitivity for sound of young people, an absolute number is defined for the sound intensity and sound pressure: 0 dB equals a power intensity of W/m^2 and a sound pressure of 2.10^{-5}Pa.

Listening conditions in recording studios require a sound pressure level of around 80 dB at the listening position with dynamic peaks of 110 dB! This means a power range of a factor 1000. If one would like to reproduce this level at home with an average loudspeaker with a sensitivity of 85 dB per watt @ 1 metre distance this would require a total amplifier power of all channels combined of more

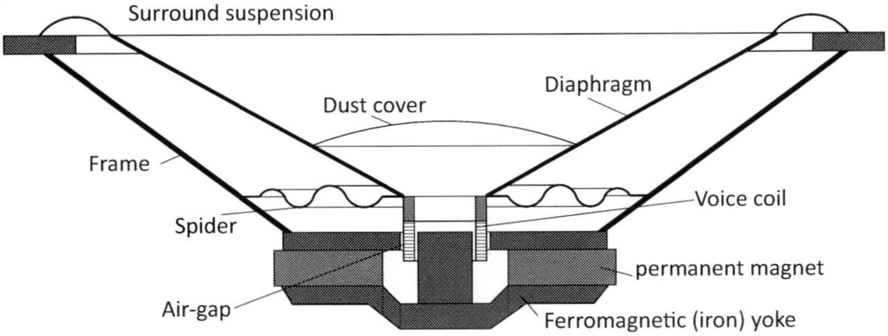

Figure 2: Cross section of an electro-dynamic loudspeaker

than 1000W because the listening position is always further away than 1 metre and the relation between the local intensity and the distance is quadratic. So twice the distance from the loudspeaker is a factor 4 reduction of power (= -6 dB).
For a Cinema the distances are even higher, requiring an installed power of several kilowatts. As will be shown this level requires also very large loudspeakers to achieve this sound intensity level at low frequencies.

Sound Power radiation from an electro-dynamic loudspeaker

Electro-dynamic loudspeakers consist of a linear Lorentz-type actuator, a membrane, often in a cone shape and a supporting frame like shown in **Figure 2**. The Lorentz-type actuator consists of a permanent magnet "stator" creating a strong magnetic field inside an air-gap and a moving coil inside that air-gap that reacts on the current in the coil by a proportional force on the membrane which in its turn pushes and pulls the air.
Because of the fact that the electro-dynamic loudspeaker was one of the first applications of Lorentz actuators this type of motor is often called a "loudspeaker motor" and the moving coil is also called a "voice coil" because that coil gives the voice to the loudspeaker. The membrane moves ideally like a stiff piston, creating pressure waves that are propagated through the air. The sound power radiated from the membrane, when moving with a sinusoidal motion, is the multiplication of the effective or root mean square (RMS) value of the velocity of the membrane with the effective value of the real (in phase) part of the force on the membrane caused by the pressure that is exerted on the membrane.

$$P_{\mathrm{a}} = v_{\mathrm{d}} F_{\mathrm{d}} = v_{\mathrm{d}} p_{\mathrm{d}} A_{\mathrm{d}} \tag{2.4}$$

where:
P_a = Acoustic power [W]
v_d = Effective value of the velocity of the membrane [m/s]
F_d = Effective value of the force on the membrane [N]
P_d = Pressure on membrane [N/m^2]
A_d = Surface of membrane [m^2]
Note that the effective (RMS) value is defined as the value of an equivalent DC value of the parame-

ter that creates the same power as the alternating version of the parameter. The effective value of a sinusoidal voltage is found by calculating the root of the mean value of the signal squared and equals:

$$V_{\mathrm{eff}} = \frac{\hat{V}}{\sqrt{2}} \tag{2.5}$$

where $\hat{V}$ = the amplitude of the voltage.
The pressure on the membrane is caused by the movement of the membrane itself with the following relation:

$$p_{\mathrm{d}} = v_{\mathrm{d}} A_{\mathrm{d}} R_{\mathrm{a}} \quad [\mathrm{N/m^2}] \tag{2.6}$$

where R_a = the real part of the acoustic impedance.
At higher frequencies this acoustic resistance is not constant due to the relation of the size and the wavelength leading to diffraction and bundling. For low frequencies it can be approximated by:

$$R_{\mathrm{a}} = \frac{\rho_0 \omega^2}{2\pi c_0} \quad [\mathrm{Ns/m^5}] \tag{2.7}$$

where:
ρ_0 = density of air [kg/m³]
ω = angular frequency of the signal [rad/s]
c_0 = velocity of sound [m/s]
With these values and equations the force by the air on the membrane can be calculated:

$$F_{\mathrm{d}} = p_{\mathrm{d}} A_{\mathrm{d}} = v_{\mathrm{d}} A_{\mathrm{d}}^2 R_{\mathrm{a}} \quad [\mathrm{N}] \tag{2.8}$$

Which can also be written as:

$$F_{\mathrm{d}} = v_{\mathrm{d}} \frac{A_{\mathrm{d}}^2 \rho_0 \omega^2}{2\pi c_0} \quad [\mathrm{N}] \tag{2.9}$$

The acoustic radiated power is equal to:

$$P_{\mathrm{a}} = v_{\mathrm{d}} F_{\mathrm{d}} = v_{\mathrm{d}}^2 \frac{A_{\mathrm{d}}^2 \rho_0 \omega^2}{2\pi c_0} \quad [\mathrm{W}] \tag{2.10}$$

Assuming a constant frequency-independent motion-amplitude of the membrane ($v_d = x_d\omega$ with x_d= the effective value of the membrane excursion) the power is written as:

$$P_{\mathrm{a}} = x_{\mathrm{d}}^2 \frac{A_{\mathrm{d}}^2 \rho_0 \omega^4}{2\pi c_0} \quad [\mathrm{W}] \tag{2.11}$$

This means that at constant displacement amplitude of the membrane, the radiated power increases proportional to frequency to the power 4. In the Bode-plot this means a slope of +2 being12 dB/octave. It also means that the radiated power in dB is proportional to the acceleration of the membrane. This also increases with frequency with a slope of 12 dB/octave when the amplitude of the displacement is kept constant as can be seen by taking the double derivative of the position:

$$x(t)=\hat{x}\sin(\omega t),\Rightarrow v(t)=\frac{\mathrm{d}x(t)}{\mathrm{d}t}=\hat{x}\omega\cos(\omega t),\Rightarrow a(t)=\frac{\mathrm{d}^2x(t)}{\mathrm{d}t^2}=-\hat{x}\omega^2\sin(\omega t) \qquad (2.12)$$

This relation of sound power with acceleration is an important property as it means that the dynamic behaviour of a loudspeaker element for low frequencies can be found by determining the acceleration as function of the frequency of a corresponding damped mass-spring system.

While the mass and the stiffness are often easily derived from the data given by the manufacturer of the loudspeaker, modeling also requires that the damping coefficient is known. Damping corresponds with the energy that is dissipated either by acoustic radiation or in the electrical circuit. The acoustic damping from the air can be derived from Equation 2.9:

$$c_{\mathrm{a}}=2\frac{A_{\mathrm{d}}^2\rho_0\omega^2}{2\pi c_0}\quad[\mathrm{Ns/m}] \qquad (2.13)$$

The factor 2 is added because for damping both sides of the membrane will give energy to the air. The energy at the backside is however of limited use when the loudspeaker element is mounted in the enclosure. It will be shown that this acoustic damping gives a very limited contribution to the total damping, indicating the low efficiency of an electro-dynamic loudspeaker.

Limitation of motional feedback to low frequencies

As mentioned before the equation for the acoustic resistance is only valid for low frequencies where the wavelength is long in respect to the diameter of the membrane. While the wavelength of sound is proportional to the velocity of sound and inversely proportional to the frequency, the wavelength for 100 Hz and below is more than 3 metres which for any commercial loudspeaker is far larger than its size.

At higher frequencies the wave front will gradually narrow into a forward directed bundle of sound which would give a higher sound pressure level (SPL) on-axis while at the same time this effect is compensated by the reduction of the radiated power because the acoustic resistance becomes independent of the frequency at these frequencies. Also the self-inductance of the actuator reduces the current through the coil at higher frequencies because a standard audio amplifier acts as a voltage-source independent of the impedance of the loudspeaker. Last but not least, at higher frequencies the membrane itself will deform creating resonance modes that impact the radiated sound. These higher-frequency phenomena can be neglected in this case-study because of the focus on the low-frequency spectrum below 80 Hz to create a THX© class subwoofer. It will be shown that the low-frequency behaviour is determined by the primary resonance of the mass of the membrane with the stiffness of the support needed to define its position. That support consists of the spider, a membrane near the coil and the rubber surround suspension at the edge of the membrane. This total stiffness creates a resonance frequency together with the mass of the membrane and the coil when

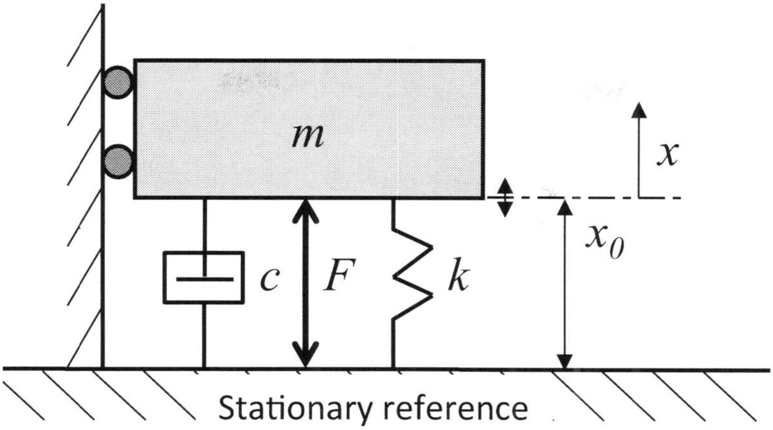

Figure 3: A damped mass-spring system with an external force stimulus.

the loudspeaker element itself is observed without an enclosure. The resulting dynamic behaviour shows an attenuation of the frequencies below this resonance frequency due to the above explained relation between acceleration amplitude, frequency and radiated power of the membrane.
Another limitation of the low frequency behaviour is the fact that both the front-side and the back-side of the membrane create sound waves that are in opposite phase because when the membrane moves to the front the pressure at the front surface will rise but on the backside the pressure will decrease. At low frequencies, when the wavelength is long, these pressure waves cancel each other. For this reason in most cases the loudspeaker is mounted in an enclosure that cancels the sound from the back.

Dynamic properties of an electro-dynamic loudspeaker
For low frequencies an electro-dynamic loudspeaker can be described in a simple model as shown in **Figure 3**. The moving body with mass, consists of the membrane and the coil. The body is suspended by the spider and the rubber roll surround, with a certain stiffness. Finally the damper, with damping coefficient c represents the loss of energy by the sound radiation, the rubber surround and the actuator-amplifier combination. With Equation 2.13 the damping due to the sound can be calculated. This damping is frequency dependent so it is necessary to calculate its level for a certain frequency. As will become clear in the actual example however, the acoustic damping is so low in respect to the other causes of damping that it is allowed to neglect the acoustic effects on mass and damping. Also the effect of the surround is small compared with the electromagnetic damping of the motor in combination with the amplifier.
For further analysis, we will use the electrical circuit of **Figure 4**. For the electrical signal source to the loudspeaker a usually applied amplifier with a voltage source output is chosen. A voltage source output means that the amplifier has a very low output impedance R_s, (in practice <100mΩ).
The electrical circuit of the loudspeaker consists of the resistance and self-inductance of the coil

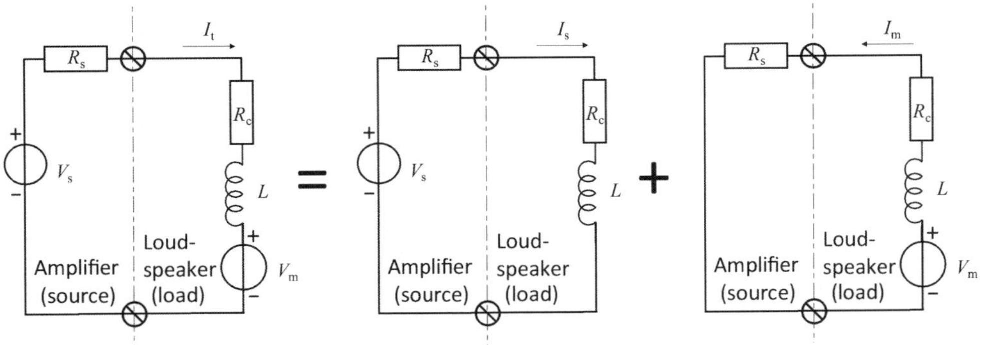

Figure 4: The electrical model of the amplifier loudspeaker combination. The total current though the loudspeaker consists of the part delivered by the voltage source of the amplifier and the part delivered by the motion of the loudspeaker.

windings in series with a voltage source with a voltage that is proportional to the velocity of the coil relative to the magnetic field.

The voltage source of the amplifier and the motor voltage are connected in series with the total impedance, causing the current (I_t) that creates the force to the moving part of the loudspeaker. The effect of both voltages can be analysed separately as their currents can be super-imposed because of the linear relationship. The low frequency also allows to neglect the self-inductance so the impedance as used in the model is resistive with value $R = R_s + R_C$. R_C is the resistance of the coil and because of the low output impedance of the amplifier it is allowed to approximate to $R \approx R_C$.

First the effect of the amplifier voltage is determined only by replacing the motor voltage by a short circuit which is allowed as the impedance of a voltage source is zero.

The force of the motor is found by using Ohm's law and the force relation of a Lorentz actuator.

$$F = BI_s\ell = \frac{BV_s\ell}{R_c} \quad [\mathrm{N}] \tag{2.14}$$

where B equals the magnetic flux density in Tesla through the coil and equals the length of the windings of the coil inside the magnetic field.

When writing the equations of motion for a damped mass-spring system [3], the frequency response function $T_{F,x}$ between force and excursion of the moving part with mass can be written as follows as function of radial frequency: $\omega = 2\pi f$

$$T_{F,x}(\omega) = \frac{x}{F} = \frac{C_s}{-\frac{\omega^2}{\omega_0^2} + 2j\zeta\frac{\omega}{\omega_0} + 1} \tag{2.15}$$

where damping ratio $\zeta = \dfrac{c}{2\sqrt{km}}$, compliance $C_s = \dfrac{1}{k}$ and resonance frequency $\omega_0 = \sqrt{\dfrac{k}{m}}$

This transfer function shows a constant value of C_s for very low frequencies ($\omega << \omega_0$). At $\omega = \omega_0$ the denominator becomes equal to the damping ratio related term which is zero when there is no damping ($\zeta = 0$), corresponding with a high resonance peak. At frequencies above the first term of the denominator is dominant, indicating an attenuation that is proportional to the frequency squared.
The acoustic response of a loudspeaker is shown to be proportional to acceleration which means that according to Equation 2.12. this transfer function has to be multiplied with ω^2 to get the requested behaviour.
Combined with Equation 2.14. the transfer function $T_{v,x}$ from voltage to acceleration becomes a func-

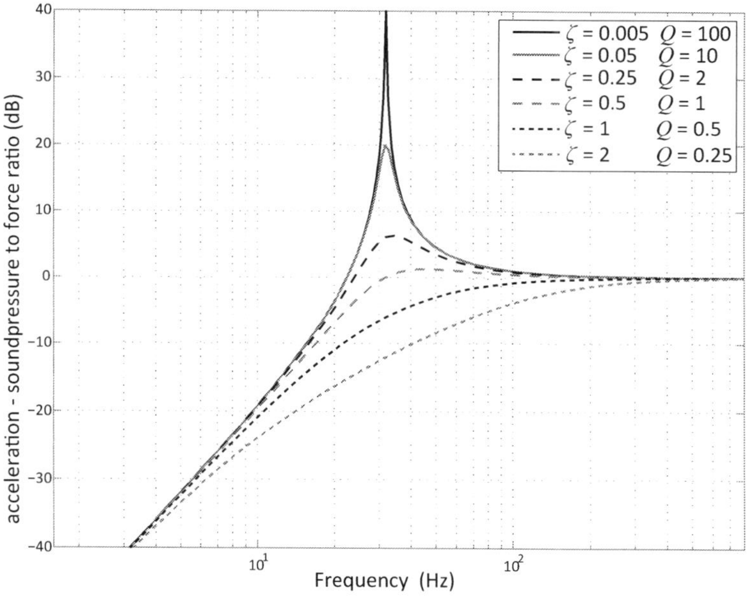

Figure 5: Bode plot of the sound response to an excitation force of an electrodynamic loudspeaker normalized to a dynamic system constant of one with different damping settings.

tion of radial frequency $\omega = 2\pi f$:

$$T_{V,\ddot{x}} = -\frac{\omega^2 x}{V_s} = -\frac{\dfrac{\omega^2 C_s B\ell}{R_c}}{-\dfrac{\omega^2}{\omega_0^2} + 2j\zeta\dfrac{\omega}{\omega_0} + 1} = -\frac{\omega^2 K_e}{-\dfrac{\omega^2}{\omega_0^2} + 2j\zeta\dfrac{\omega}{\omega_0} + 1} \tag{2.16}$$

with the dynamic system constant $K_{\mathrm{e}} = \dfrac{C_{\mathrm{s}}B\ell}{R_{\mathrm{c}}}$

This is shown in an amplitude Bode plot normalised to $K_e = 1$ in **Figure 5.** This results in a flat response above the resonance frequency and a decreasing response with a slope of +2 at lower frequencies. In this graph both the damping ratio ζ and the quality factor *Q* are mentioned as Q is mostly used in

loudspeaker systems. These terms relate quite straightforward as: $Q = \dfrac{1}{2\zeta}$ and $\zeta = \dfrac{1}{2Q}$.

.

The second step is to determine the force by the current fraction (I_m) that is induced into the amplifier by the motion voltage of the moving coil. The output of the amplifier can also be approximated as a low impedance circuit loading the motion voltage. The motion related current value is determined by the motion voltage and R_c and causes a force that counteracts the movement which means that it acts like a damper.
The motion induced voltage of a moving coil with velocityin a magnetic field equals:

$$V_{\mathrm{m}} = B\ell v \quad [\mathrm{V}] \tag{2.17}$$

The resulting current equals according to Ohm's law:

$$I_{\mathrm{m}} = \frac{V_{\mathrm{m}}}{R_{\mathrm{c}}} = \frac{B\ell v}{R_{\mathrm{c}}} \quad [\mathrm{A}] \tag{2.18}$$

As the current will flow in the same magnetic field, a damping Lorentz force will occur:

$$F_{\mathrm{d}} = B\ell I_{\mathrm{m}} \quad [\mathrm{N}] \tag{2.19}$$

With the previously derived value for I_m this gives:

$$F_{\mathrm{d}} = \frac{(B\ell)^2 v}{R_{\mathrm{c}}} \quad [\mathrm{N}] \tag{2.20}$$

So the damping coefficient *c* is:

$$c = \frac{F_{\mathrm{d}}}{v} = \frac{(B\ell)^2}{R_{\mathrm{c}}} \quad [\mathrm{Ns/m}] \tag{2.21}$$

Combining this damping coefficient *c* with the spring-stiffness and mass gives the damping ratio ζ, which is directly related to the electrical Q_{es} factor given in the data sheets. As will be demonstrated in the practical example of the following section the value of the damping coefficient of the applied loudspeaker element is in the order of 40 Ns/m.

As said before, the other damping factors are caused by the acoustical damping and the damping of spider and surround membrane.

For the acoustical damping we use Equation 2.13 with some practical values for a loudspeaker to get a feel for the magnitude.
With

$$A_{\mathrm{d}} = 500{\cdot}10^{-4}\ \ \mathrm{m}^2, f = 40\ \mathrm{Hz} \Rightarrow \omega = 2\pi f \approx 250\ \mathrm{rad/s},\ \rho_0 = 1,2\ \mathrm{kg/m}^3,\ \text{and}\ c_0 = 340\ \mathrm{m/s}$$

we get:

$$c_{\mathrm{a}} = 2\frac{A_{\mathrm{d}}^2 \rho_0 \omega^2}{2\pi c_0} = 2\frac{(5{\cdot}10^{-2})^2{\cdot}1.2{\cdot}250^2}{2\pi{\cdot}340} = 0.18 \quad [\mathrm{Ns/m}] \tag{2.22}$$

This is obviously such a small value in respect to the electromagnetic damping of 40 Ns/m that it can be neglected and this conclusion points again to the bad energy transfer efficiency of an electrodynamic loudspeaker.
As a next step a practical loudspeaker element is used to illustrate these theoretical derivations. The chosen example loudspeaker is specially designed for use in a subwoofer system and the derived equations will be matched with the given data to determine different characteristics that will be used in the design of the full system.
Note that in many publications the term loudspeaker is often used for both the loudspeaker element (also called the *driver*) and the total system consisting of different loudspeaker elements, the enclosure and other things like filters. Although it is tried to be avoided in this paper sometimes the term loudspeaker is used for the driver element for reason of readability.

3 The design of an active THX© subwoofer

The targeted specifications

The system is designed according to the following THX© specification of Lucasfilm© for a large room:

1 THX certified frequency response from 20 – 80 Hz with a maximum deviation of +/-3 dB.
2 Size of the room is 5 metre wide and 10 metre deep.
3 Maximum (peak) sound intensity in the sweet spot of the room @ 5 metre distance:
   - 96 dB @ 20 Hz .
   - 106 dB @ 40 – 80 Hz.
4 Aperiodic response around the main resonance frequency (*Q = 0.7*), this means no resonances nor overshoot.
5 Size of the box less than 0.1m³.
6 No external vibrations exerted to the floor to avoid unwanted sounds from contact resonances.
7 Cost should be less than €1000 on material, including power amplifier.

The specified sound levels are based on the fact that high levels of very low frequencies require extremely large-sized loudspeaker membranes while statistically these frequencies are never as strong as frequencies around 1 kHz.
As will be shown from the following sections this still will require large loudspeakers with very strong

amplifiers.

Properties of the loudspeaker element

The used loudspeaker element is made by Peerless and its main data are given in **Figure 6.** This type is selected because of its aluminum cone which remains acting quite deterministic as a solid piston until around 1 kHz which prevents strange behaviour in the frequency range of interest. Also the size is chosen as large as possible in order to realise the required sound intensity of 106 dB. Important data for the design are the DC resistance, the resonance frequency, the force factor, the moving mass, the suspension compliance or stiffness, the motor damping and the piston area.

The loudspeaker has a long excursion range while having a longer voice coil (32 mm) than the air gap (8mm). This means that with a maximum motion amplitude of the membrane of +/- 12 mm the permanent magnet flux on the coil will remain approximately unchanged which otherwise would cause harmonic distortions due to non-linearity. The efficiency is of course lower than with equal lengths of coil and magnet gap. The stiffness from the surround and spider is calculated from the resonance frequency and compared with the compliance that should equal *1/k*.

The resonance frequency is $\omega_0 = 2\pi f = 2\pi \cdot 19.1 = 120 rad/s$. With $m = 130.6 \cdot 10^{-3}$ and $k = m\omega_0^2$ the stiffness

Specs:

Electrical and mechanical Data			
Nominal impedance	Zn	4	ohm
Minimum impedance	Zmin	3	ohm
Maximum impedance	Zo	65.7	ohm
DC resistance	Re	2.6	ohm
Voice coil inductance	Le	1.6	mH
Voice coil diameter		51	mm
Voice coil height		32.6	mm
Voice coil layers		4	
Height of the gap		8	mm
Linear excursion +/-		13	mm
Total useful flux		2.3	mWb
Diameter of magnet		147	mm
Height of magnet		35	mm
Weight of magnet		2.2	Kg

T-S Parameters			
Resonance Frequency	fs	19.1	Hz
Mechanical Q factor	Qms	9.29	
Electrical Q factor	Qes	0.38	
Total Q factor	Qts	0.37	
Force factor	Bl	10.3	Tm
Mechanical resistance	Rms	1.69	Kg/s
Moving mass	Mms	130.6	g
Suspension compliance	Cms	0.53	mm/N
Effective cone diameter	D	24.4	cm
Effective piston area	Sd	466	cm^2
Equivalent volume	Vas	159	ltrs
Sensitivity		91.2	dB

www.tymphany.com

Figure 6: Peerless 30 cm woofer (a) with data (b).

becomes k = $1.87 \cdot 10^3$ N/m. This is almost equal to the inverse of the given compliance of $0.53 \cdot 10^{-3}$ m/N. The damping coefficient is derived with the previously found relation:

$$c = \frac{(BL)^2}{R} \approx 41 \quad \text{Ns/m} \tag{3.1}$$

The resulting damping ratio is:,

$$:\zeta = \frac{c}{2\sqrt{km}} = \frac{41}{2\sqrt{1.87 \cdot 131}} = 1.34,$$

with a correspondingfactor of:

$$Q = \frac{1}{2\zeta} = 0.37$$

In the given characteristics the value for the electrical Q_{ES} factor is 0.38 which is within a reasonable tolerance for these kind of systems ($< 1dB \approx 10\%$).

The Q factor for the acoustical damping as found in the previous chapter would result in a value of around 70 indicating again the low efficiency. The given mechanical *Q* factor (Q_{ms}) of 9.29 is deter-

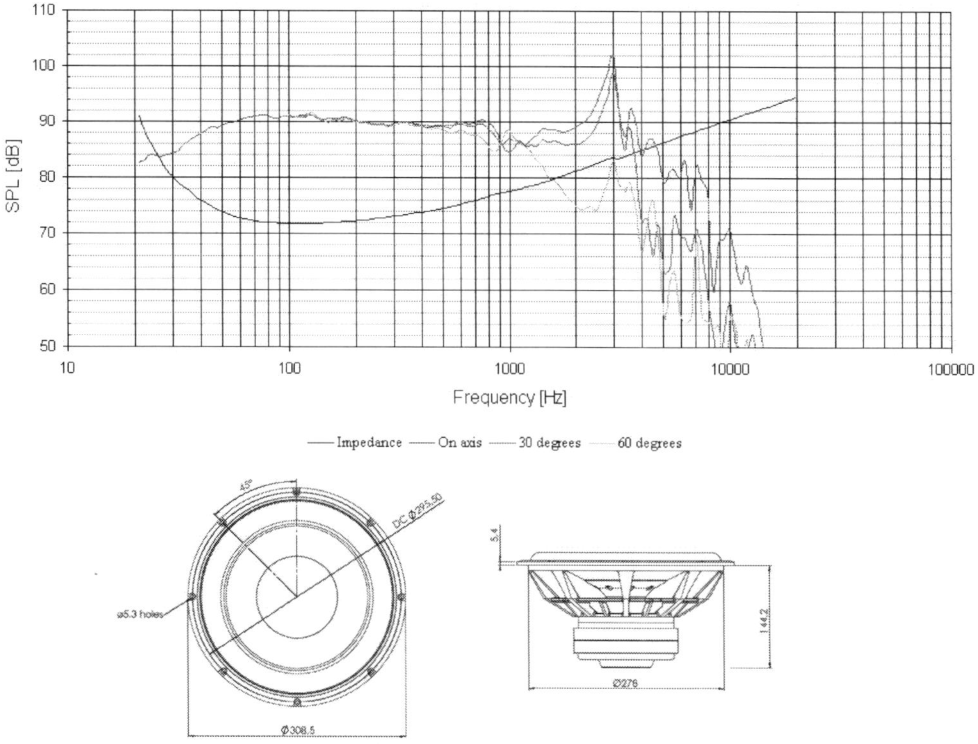

Figure 7: Response data (a) and mechanical drawing (b) of the used loudspeaker element.

mined by the spider, the surround and the air. This means that the mechanical guiding takes more energy than the acoustic damping although it determines still a very limited portion of the total damping. With a total Q of 0.37 this system represents an over damped behaviour conform the given measured response of the loudspeaker as shown in **Figure 7**. This response is normally measured mounted on an 'infinite' baffle that isolates the sound of the backside from the sound of the front side of the membrane without increasing the stiffness. The measuring conditions are an anechoic room so without reverberations. This situation is not representative for a real situation with the loudspeaker build in an enclosure and it will become clear why the manufacturer chose the parameters such that the damping is so high in the infinite baffle case.

Figure 7 also shows the impedance of the loudspeaker in equivalent ohms. In principle this is not really correct as it is a complex impedance with phase effects due to the self-inductance. At around 100 Hz the impedance is real and the current is in phase with the input voltage with a value of 6 Ohm just above the DC resistance of the coil (5.2 Ohm).

At higher frequencies the self-inductance increases the impedance while at frequencies below 100 Hz the motion voltage of the motor starts to reduce the current in the system to reach a peak at the resonance frequency where the velocity and motion induced voltage are in phase with the amplifier voltage.

Consequences of the efficiency on the realizable specifications

When looking back at the specifications it is shown that, when using one loudspeaker element, 7.5 mW acoustical power is obtained at 2.84 Watt input power. 106 dB @ 1 metre distance would require +15 dB = approximately 30 times this power level, hence 90 W electrical power. At a distance of 5 metre this same sound pressure level would however require $5^2=25$ times more electrical power, which is more than 2 kW! Fortunately the room walls help a bit at these very low frequencies as the sound cannot escape beyond the walls and is reflected at surfaces near the source, which act as an additional membrane! When the loudspeaker is placed against the wall, the floor and the wall each provide approximately a factor two in power gain (3+3=6 dB) which is a factor 4 in power, resulting in a less extreme 500 Watt of required electrical power. Placing in a corner would even give an additional 3 dB gain.

The power related specifications of the subwoofer element are not presented in the given data but a similar element from the same manufacturer with a paper cone gives a maximum sustainable peak power per element of 350 Watt and a continuous power of 150 Watt. To increase the robustness of the system and create a balance of forces to avoid external vibrations two elements are choses which additionally reduces the necessary electrical power with a factor two based on the corresponding efficiency improvement by the double membrane surface. The resulting necessary amplifier power of around 250W total is no problem for a state-of-the-art switch-mode class-D amplifier like the one used for this example from Hypex.

A final check is necessary for the maximum cone excursion. Equation 3.2 showed that an effective value of the motion of the membrane of 0.22 mm is achieved at 100 Hz with 2.84 W input power. This equals a maximum amplitude of $0{,}22 \cdot \sqrt{2} \approx 0.3$ mm.

125 Watt input power per element would increase this amplitude, proportional to the square root of the power into:

$$x_{\mathrm{d}} = 0.3{\cdot}10^{-3}\sqrt{\frac{125}{2.84}} \approx 2{\cdot}10^{-3} \quad [\mathrm{m}] \tag{3.2}$$

This amplitude would become 12.5 mm at 40 Hz, because the amplitude increases with the frequency squared (factor 6.25). This value is the maximum for this loudspeaker element and below 40 Hz the maximum power will need to be decreased by 12 dB per octave to limit the maximum excursion magnitude. This is a 12 dB reduction for 20 Hz which is just acceptable to accommodate the specification of 96 dB@ 20 Hz.
Adding two more loudspeakers could improve this performance but the cost (now already approximately €400 excluding amplifier) would not allow that.

Mounting the loudspeaker elements in a closed box
A loudspeaker element mounted in an enclosure (loudspeaker-box) will perceive an additional spring stiffness that adds to the support stiffness and as a result the resonance frequency will increase while the damping decreases.
This will be analysed in the following section for a closed-box configuration as shown in **Figure 8**, the enclosure designed for two loudspeaker elements. It uses a special configuration where both loudspeaker elements are mounted opposite each other inside the enclosure. This guarantees a maximum acoustic coupling for the efficiency enhancement and a balanced motion in the closed box configuration reducing external vibrations. Furthermore the membranes are better protected from external damage by rough handling.
The internal volume of the enclosure is approximately 60 litres. This type of enclosure is only allowed for frequencies well below the wavelength that is defined by the size of the cavity between the loudspeakers.

At higher frequencies the sharp edges would result in diffraction phenomena creating irregularities in the radiated frequency spectrum. A benefit of the shape of the enclosure with low frequencies is the possibility to place it in the corner of a room with the opening directed to the corner as this would slightly increase the efficiency by an optimal coupling to the walls. One might say that it creates an acoustic horn from the room geometry.
The membranes are actuated opposite to each other when both loudspeaker elements are connected in phase to an amplifier. The symmetrical design allows the behaviour to be modelled as two equal but mirrored enclosures of half the size. On the symmetry-axis inside the enclosure the molecules of the air spring will not move anymore.
To calculate the coupling stiffness of the air that is enclosed between the membranes, the data are taken from the surface area of the membrane A_d of one loudspeaker element and by applying the simple gas law $PV = C$ the force is calculated as a function of the displacement. With a displacement x_d the displaced volume V_d equals x_dA_d.

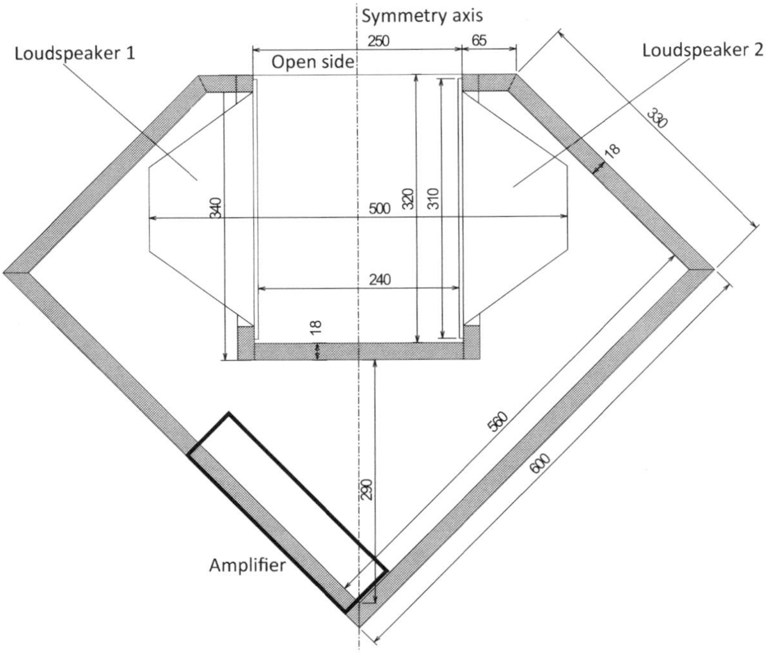

Figure 8: Drawing (a) and picture (b) of the enclosure during assembly, before the top cover is placed.

The displaced volume will cause a relative pressure change to the environmental pressure P_{env} that is calculated from the gas law and the relative change of the volume:

$$P = P_{\text{env}} \frac{V_{\text{d}}}{V_{\text{en}}} = P_{\text{env}} \frac{x_{\text{d}} A_{\text{d}}}{V_{\text{en}}} \quad [\text{N/m}^2]$$

The force on the membrane equals: $F = PA_{\text{d}} = P_{\text{env}} \frac{x_{\text{d}} A_{\text{d}}^2}{V_{\text{en}}} \quad [\text{N}]$

With the surface area of the membrane $A_d = 0.047$ m², an air pressure of 10^5Pa and a volume of the enclosure $V_{en} = 0.06$ m², the total stiffness of the spring due to the enclosed air becomes:

$$k_{\text{e}} = \frac{F}{x_{\text{d}}} = \frac{P_{\text{env}} A_{\text{d}}^2}{V_{\text{en}}} = \frac{10^5 \cdot (0.047)^2}{0.06} \approx 3.6 \cdot 10^3 \quad [\text{N/m}]$$

The compliance ($1/k$) of this spring is equally divided between the two masses. This means that each membrane will be coupled to a spring with twice the stiffness ($k_a = k_e = 7.3 \cdot 10^3$ [N/m]). Adding the stiffness of the cone suspension results in the total stiffness per membrane:

$$k_{\text{m}} = (1.86 + 7.3) \cdot 10^3 \approx 9.2 \cdot 10^3 \quad [\text{N/m}]$$

which gives a resonance frequency of:

$$f_0 = \frac{\omega}{2\pi} = \frac{1}{2\pi} \sqrt{\frac{k}{m}} = 42 \quad [\text{Hz}]$$

The damping coefficient c, when applying a voltage source amplifier, was calculated with Equation 3.1 to be 41 Ns/m, resulting in a damping ratio of: $\zeta = \frac{c}{2\sqrt{km}} = \frac{41}{2\sqrt{9.2 \cdot 131}} = 0.6$

with a corresponding $Q = \frac{1}{2\zeta} = 0.83$

It can be concluded that the resonance frequency is higher than in the un-mounted situation and the damping ratio is increased to a more optimal value which is the reason why the manufacturer chose the initial values of the parameters. A value of $Q = \zeta \approx 0.7$ would be the most optimal (aperiodic) value without any resonance nor excessive overshoot in the step response. The step response is a measure for the character and should show as little overshoot as possible to prevent "boom-box" effects. **Figure 9** shows that the step response with $Q = 0.83$ is indeed not without some overshoot but what is even worse is that the lowest -3 dB frequency limit is a full octave above the targeted THX specification of -3dB at 20 Hz.

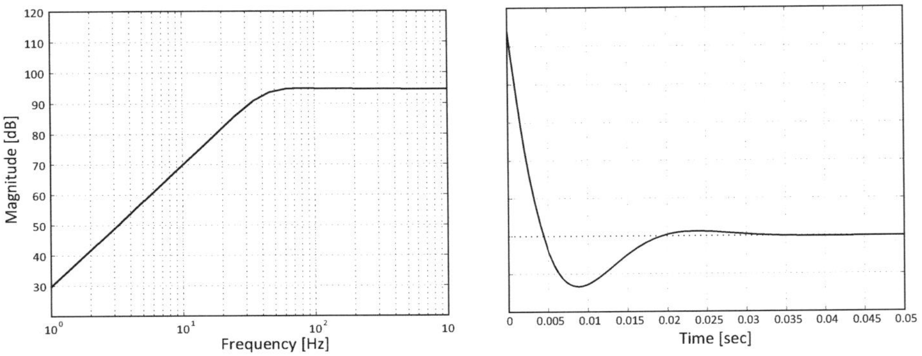

Figure 9: Frequency and step response of the closed-box system when driven with a voltage source amplifier.

To improve this dynamic behaviour there are several options:

- Compensate the characteristics with an electronic filter with an inverse transfer function of the loudspeaker.
- Measure the sound and use acoustical feedback.
- Measure the motion of the membrane and use "motional feedback"'.
- Use a different amplifier.

The use of an inverse filter is the most used option in commercially available subwoofers [4] although it is hardly possible to do this accurately especially with situations where the damping is low. The exact tuning of the dynamic response including overshoot is for that reason often less than acceptable.
An improvement of this principle can be obtained using digital electronics to create a more exact inverse transfer function by means of a Digital Signal Processor (DSP) with built in amplifiers dedicated to each loudspeaker. One early example of such a system was the "digital loudspeaker" of Philips that was unfortunately so badly designed in other aspects like an un-damped very thin enclosure wall and the cheapest choice of amplifiers and loudspeaker elements that it utterly failed in the market (again, what a waste!). The unsolved problem with this approach is that it is purely based on feed-forward compensation while it will force the loudspeaker to very high uncontrolled excursions at very low frequencies into the non-linear range of the surround-spring and the Lorentz actuator with a resulting severe distortion that can only be partly compensated in the DSP.
Another method is to measure the sound itself by a microphone and use that information as a feedback signal for an actively controlled system. This approach is in practice limited by the delay between the loudspeaker and the microphone and by the room reflections that are also delayed and detected. Although this principle can be used with adequate filters to suppress room reverberations, these delays would cause the system to become unstable resulting in a howling sound. Furthermore it would be limited to very low frequencies where the wavelength is long as at higher frequencies it is impossible to create a coherent wave front that is not impacted by these reverberations.
As mentioned in the introduction, a better method is to use Motional feedback by directly measuring the movement of the membrane and use that in an actively controlled system.

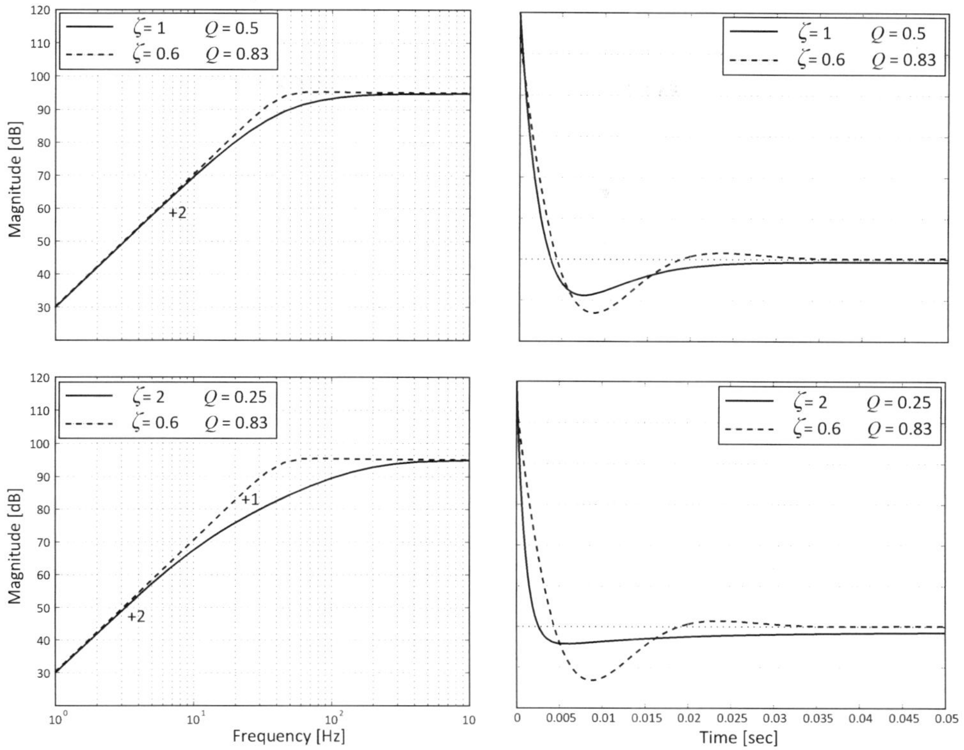

Figure 10: Frequency and step response of the closed-box system at different damping levels compared to the original damping factor when driving with a voltage-source amplifier.

Motional feedback by active control of the damping

When presenting the dynamic properties of an electro-dynamic loudspeaker, it was shown that the resonance of a loudspeaker element is effectively damped by short circuiting the motion induced voltage by the amplifier. This phenomenon is in fact a very simple way of proportional velocity feedback as the velocity is slowed down by a force generated by the motion voltage proportional to that same velocity. This damping-inducing feedback phenomenon is determined by the total resistance of the electrical loop where the motion voltage of the Lorentz actuator creates the current over this loop-resistance causing the damping force in the actuator. By reducing the total resistance *below the DC resistance of the loudspeaker itself* very high levels of damping would be obtained which also would reduce the distortion by the velocity feedback and completely cancel the overshoot. **Figure 10** shows the beneficial effect of an increase of damping to $\zeta = 1$ or 2 ($Q = 0.5$ or 0.25).

The resonance peak is completely vanished but as a drawback the down-slope starts at a higher frequency albeit more gradual. In fact the system starts to act like a series of two first-order high-pass filters. At $\zeta = 2$ ($Q = 0.25$) the slope angle in the frequency range of interest between 20 and 80 Hz is

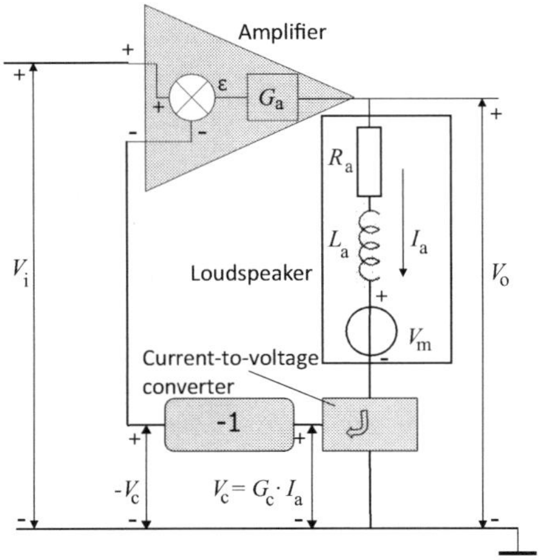

Figure 11: A positive current feedback will create an amplifier with a negative source impedance to the load.

about 20 dB per decade (+1) like a first-order high-pass filter so it should be possible to compensate that quite easily with a first-order filter.

But now the question arises how the total loop-resistance can be reduced to below the resistance of the moving-coil. With a voltage amplifier this total resistance equals the coil resistance and the connecting wires only. As long as the resistance of these wires is less than 5 % of the coil resistance their effect on the damping can be neglected in spite of all the stories about loudspeaker cables. But it is possible to create a far more significant effect by a special trick in the amplifier that causes it to supply more voltage when the load demands more current. This means that the amplifier creates a negative resistive source impedance for the load which effectively reduces the total resistance.

The working principle is explained as follows: From electronics theory [3] it is known that a transconductance amplifier uses negative current-feedback to create a very high output impedance. The negative feedback-control keeps the output current at a value that is only determined by the setpoint, independent of the output voltage. From this fact - that a negative current-feedback increases the output impedance - one can conclude that the reverse situation, positive current feedback, would decrease the output impedance. This is visualised in a more schematic way shown in **Figure 11**. The current through the load is measured by a current-to-voltage converter with a negative sign to compensate for the negative sign at the input of the amplifier, creating positive current feedback. A higher output current will then result in an increased output voltage.

From control theory [3] it is known that the closed-loop gain of an amplifier with normal negative feedback can be written as follows:

$$G_{\mathrm{cl}} = \frac{V_{\mathrm{o}}}{V_{\mathrm{i}}} = \frac{G_{\mathrm{a}}}{1 + G_{\mathrm{a}} G_{\mathrm{fb}}} \tag{3.3}$$

With G_a being the gain of the amplifier in the forward path and G_{fb} the gain of the feedback path. This feedback gain can be derived from the gain of the current-to-voltage converter by calculating the current I_a as function of the output voltage V_o using Ohm's law:

$$I_{\mathrm{a}} = \frac{V}{Z} = \frac{V_{\mathrm{o}} - V_{\mathrm{m}}}{R_{\mathrm{a}} + j\omega L_{\mathrm{a}}} \tag{3.4}$$

The self-inductance l_a can be neglected because of the low frequencies. The motion induced voltage source V_m has no relation with the effect of the output voltage of the amplifier on the current and can be replaced by a short circuit to determine the feedback gain:

$$G_{fb} = \frac{V_c}{V_o} = -\frac{G_c I_a}{V_o} = -\frac{G_c}{R_a} \tag{3.5}$$

Using this result in Equation 3.4 the following closed-loop gain is obtained:

$$G_{cl} = \frac{V_o}{V_i} = \frac{G_a}{1 - G_a \dfrac{G_c}{R_a}} \tag{3.7}$$

The term in the denominator would become 1 when $G_c = 0$ and the closed-loop gain is then equal to the original gain of the amplifier. The closed-loop gain would become infinite when $G_c = R_a/G_a$. Between these values the system is stable and the closed-loop gain is increased relative to the open-loop amplifier gain with a factor $1/(1\text{-}G_aG_c/R_a)$.

In the last part of this paper this effect is used in the realisation of a motional-feedback loudspeaker to create a negative output impedance. The value of the output impedance can be calculated by the fact that it represents the change of the output voltage due to the output current. In case the impedance is resistive and positive the output voltage decreases (dV_o) with an increase (dI_a) of load current, $dV_o = -R_o dI_a$.

In the situation of Figure 11 an increase in load current will create an *increase* in output voltage as there is twice the minus term (-1×-1 = 1) in the loop so $dV_o = dI_aG_cG_a$, which automatically leads to the conclusion that $R_o = -G_cG_a$.

By adjusting both gains the negative output impedance can be set, such that the total loop impedance is tuned to a suitable value to achieve the required level of damping.

Practical realisation and results

Figure 12 shows the electronic circuit of the test set-up. The first operational amplifier buffers the input and creates the first-order bass-boost compensation filter. The second operational amplifier is an inverting amplifier where the input voltages and current feedback signal are added. A Hypex switched-mode power amplifier in bridged mode is used. This amplifier has a voltage source output with almost zero output impedance and ample power reserve. A positive current feedback loop is added to create the negative output impedance. The current is measured by a current to voltage converter of the company LEM. It consists of a ferromagnetic ring around the current-carrying wire with an additional compensation winding wound around the ring. The related electronics measure the induced magnetic field in the ring and enables the measurement of current without losses by compensating this field with the compensation windings. A series resistor can be used as a more affordable alternative, when applying a single ended output power amplifier, according to Figure 11.

The coils of the two loudspeakers are connected in series which is allowed as they share the same dynamics, which means that all voltages and currents are in phase. The resulting current and the re-

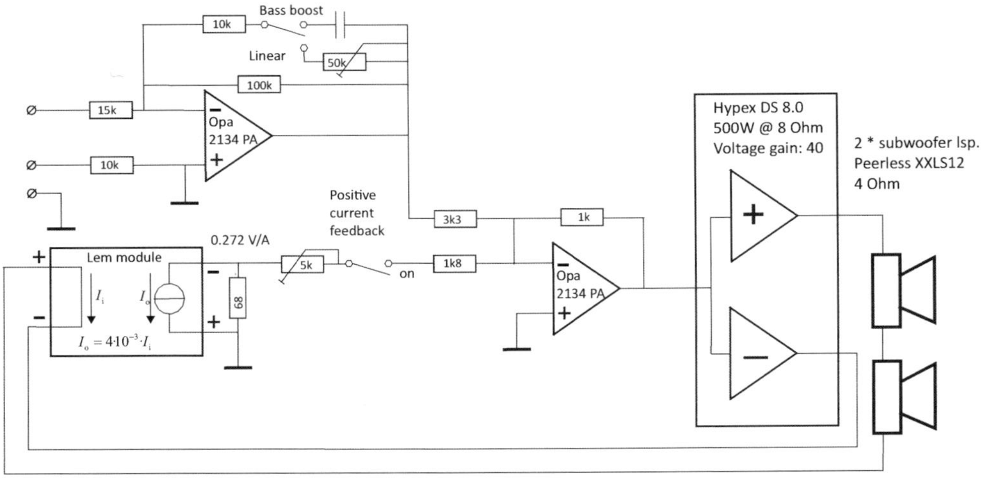

Figure 12: Electronic circuit of the test set-up showing the main components and some tunable elements. With the upper switch the compensation circuit can be bypassed while the corresponding potentiometer can be adjusted to limit the gain difference at switching.

sulting damping force remain the same as in the situation with one loudspeaker element because both the motion voltage and the total resistance are doubled. Neglecting the self-inductance, the resulting total resistance is 5.2 Ohm and with a voltage-source amplifier the previously calculated value of the damping ratio $\zeta = 0.6$ would be obtained. This value has to be increased with at least a factor of 2 or even more when possible to get the maximum controlled motion. To achieve this high damping ratio the maximum total resistance must be reduced to $5.2{\cdot}0.6/2 = 1.56$ Ohm.

This means that the amplifier must have a negative output impedance of approximately -3.6 Ohm. This is set with the 5 kOhm potentiometer. To determine the approximate setting, Equation $R_o = -G_cG_a$ is used with $G_a = 40$ and with the potentiometer set at zero. In that situation G_c equals the total gain of the path from the current via the current-to-voltage converter and the difference amplifier to the input of the amplifier being $G_c = -0.272{\cdot}8 \approx -0.15$. This results in a negative output resistance of 6 Ohm which is just too much but by increasing the value of the potentiometer, the negative impedance can be reduced until the right value is obtained. In practice the self-inductance still shows to have a little bit of influence by increasing the impedance of the loudspeakers as the system is still just stable at a value of 6 for the negative output impedance.

The first operational amplifier acts as the required first-order "bass-boost" filter. Below 160 Hz the gain will increase until 16 Hz where it becomes flat again. This compensates the first order +1 slope effect of the negative output impedance nicely as can be seen in **Figure 13** where the response is simulated. A flat response with a low frequency -3 dB bandwidth limitation at 20 Hz has been realised for the total system as was verified with measurements on the cone excursions. Unfortunately,

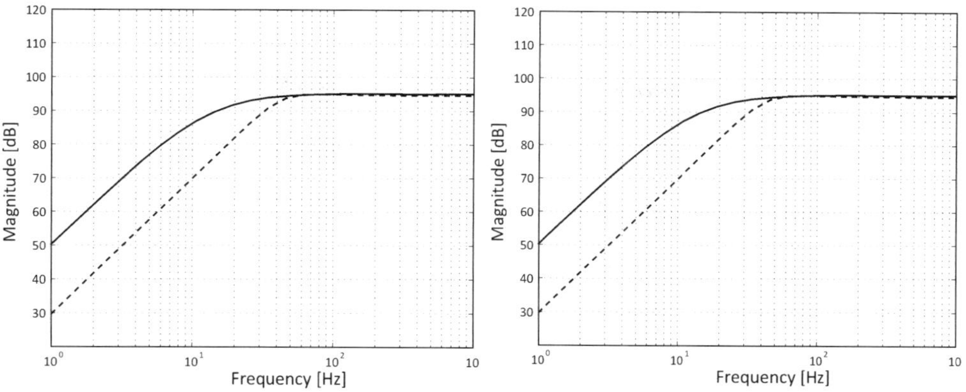

Figure 13: Response of the closed-box when driven with an amplifier with a negative output impedance compared with the passive uncontrolled response. The high damping with a first-order compensation for the low frequency roll off creates a system with a -3 dB bandwidth at 20 Hz.

the lack of an anechoic room of sufficient size did not enable reliable sound field measurements but the subjective sound impression does not contradict the found results.

It is necessary to be aware of the additional amplification (gain) of this circuit caused by the positive current feedback.
When using Equation 3.7 an almost infinite gain would be obtained as soon as the negative output impedance equals R_a. When the system is tuned close to this level, the total gain becomes very sensitive for resistance and gain variations of the elements in the feedback loop which is also a reason not to tune the system too critical. For instance a damping ratio of 2 with the corresponding output impedance of 3.6 Ohm requires $G_c = -0.15 \cdot 3.6/6 = 0.09$. The gain of the system without positive current feedback would increase with a factor $1/(1-G_aG_c/R_a)$. Substituting this data shows 2.5 times the normal gain. This increased gain has to be taken into account when balancing the output of the subwoofer with the other loudspeaker systems in the surround set-up.

Some additional remarks
The presented system works and fulfills almost all of the required specifications however one other aspect still influences the behaviour and that is the effect of heating of the actuator coils at high power levels which will increase the resistance of the coil windings. This higher resistance gives a reduced damping and consequently a rise in the response around 40 Hz. In practice this effect is hardly noticeable and remains below a 3 dB change in the response. One might tune the system such that it is optimal in an average heated situation as long as the system does not become unstable at lower temperatures. A solution could be found in more advanced control strategies with real-time system identification and automatic tuning of the parameters which is beyond the scope of this case study but will be part of a research study at the university.

Furthermore, it can be reasoned that the intermodulation distortion due to the non-linear position-dependent gain of the Lorentz actuator is reduced. At large low-frequency excursions the force factor of the actuator reduces with a proportional reduction of the output of higher frequencies at these maximum excursion peaks. Without active control this magnitude modulation of the higher frequency by the lower frequency (intermodulation) would be audible. With active feedback the motion induced voltage will also reduce at the outer excursions, giving a reduced velocity feedback and consequently an increased signal magnitude of the higher frequencies, partly compensating the modulation. This is especially at frequencies around 40 Hz where the damping effect of the motion voltage is most effective as was shown in Figure 10.

Last but not least a limiter would be needed to protect the membranes from damage due to extreme motions at subsonic frequencies and to prevent the amplifier from severe clipping caused by the compensation network. This would require a dynamic equalizing filter where the maximum signal amplitude of frequencies below 50 Hz is limited.

References

[1] JWS Rayleigh a.o., The Theory of sound Vol 1 and 2 (orig.1877), Dover publications (1945).

[2] A.N. Thiele, Loudspeakers in vented boxes: Part II, Proceedings of the IRE Australia, vol 22, pp 487-508 (1961).

[3] Robert Munnig Schmidt et all, The design of High-Performance Mechatronics, Delft University Press (2011).

[4] W. Marshall Leach jr., Active equalization of closed-box loudspeaker systems, Journal of the AES (Jan. 1980).

Loudspeaker efficiency – an Intermezzo

Loudspeaker elements are specified to work with a voltage-controlled amplifier and as often more elements are mounted together in one enclosure to cover different frequency ranges these loudspeaker elements have to match in sound intensity (power). In order to achieve this matching the radiated sound level of a loudspeaker (driver) element is measured with a standard signal. This standard signal is the AC voltage level that would result in an electrical power of 1 Watt in a 8 Ohm pure resistance. The corresponding voltage has an effective value of 2.83 V RMS.

To get a feel of the efficiency an approximate calculation of the radiated power is done based on equations and this is compared with the given output of 91 dB at 1 metre with this standard voltage. The situation at 100 Hz is taken because then the impedance is real and the current can be easily determined. First the amplitude is calculated from the current induced by the input voltage. Note that all values are effective values (RMS) so the sine terms can be omitted.

The current induced by the input voltage equals: $I = \frac{V_i}{R} = \frac{2.83}{2.6} = 1.09 \quad [\mathrm{A}]$

This current causes a force of: $F = BI\ell = 10.3 \cdot 1.09 = 11.2 \quad [\mathrm{N}]$

At 100 Hz the frequency response is mainly determined by the mass and the acceleration will equal:

$$a = \frac{F}{m} = x\omega^2 \quad [\mathrm{m/s^2}]$$

This means that the excursion of the membrane will be:

$$x = \frac{F}{m\omega^2} = \frac{10}{0.130(628)^2} \approx 2.2 \cdot 10^{-4} \quad [\mathrm{m}] \qquad (3.6)$$

When using the numbers from the table Equation 2.11 gives the radiated power:

$$P_a = (2.2 \cdot 10^{-4})^2 \frac{(466 \cdot 10^{-4})^2 \cdot 1.2 \cdot (628)^4}{2\pi \cdot 340} = 9 \cdot 10^{-3} \quad [\mathrm{W}]$$

This result must be compared with the given efficiency of 91 dB which is 10^{-12} aboveW/m^2 (0dB) which is approximately $1.2 \cdot 10^{-3}$ W/m^2. This value is measured at a distance of 1 metre on an infinite baffle and at this frequency the sound radiates in one hemisphere with a radius of 1 metre so the surface is equal to half a sphere $S = 2\pi P^3 = 6.3$ [m^2].

This means that the radiated power that corresponds with the given measurement data of the loudspeaker equals $7.5 \cdot 10^{-3}$ W which is only a difference of 1.6 dB less than the calculated value. This difference might very well be due to the internal damping of the anechoic room and other factors not encountered in the calculation.

This all makes it very clear that in principle loudspeakers are not well fit for this era of sustainability and green designs when considering that 1 Amp at 2.84 V means 2.84 Watt electrical input giving less than 10 mW acoustical output power!
Besides the limited motor efficiency due to the longer coil than the air gap, the main cause for this lack of efficiency is the loose coupling of the membrane to the air, causing the membrane to move along its mass line where force and velocity are 90 degrees out of phase. Fortunately sound has often a very high dynamic range where the average power is often more than 30 dB below the peak level which is a factor 1000 in power, so the energy problems of Earth will not be solved by increasing this efficiency. In case high power amplifiers are not practical this efficiency can be improved by using a horn. This acts as an acoustic transformer that creates a high impedance at the membrane giving a high pressure with corresponding high power. Horns are very difficult to design and their size is proportional to the longest wavelength of the radiated sound. For very low frequencies their size would hence become impractical.
When searching for improved efficiency at low frequencies it is interesting to note that Equation 2.11 shows that by increasing the surface of the membrane with a factor 2, a factor 4 more power is obtained if all other parameters are kept equal. That could be done by increasing the diameter of the membrane but then the mass would increase also which would affect the amplitude. By taking a second loudspeaker that is driven by the same voltage, the amplitude is not changed and 4 times the output power is obtained at 2 times the input power. This hint towards a "Perpetuum Mobile" is only true as long as the pressure that is caused by the movement is not influencing the amplitude of the movement itself, which is the case in this situation due to the loose coupling between air and membrane. This also means that the pressure of both loudspeaker elements adds together and the result works on both membranes, which is the explanation for the gained efficiency. This effect is like so many phenomena only true when the distance between the loudspeakers is smaller than the wavelength as otherwise directional effects due to interference would spoil the linear addition of the pressure levels. This also means that, when using two subwoofers in the same room, one should reduce the low-frequency input signal with 3dB to compensate the factor two in radiated power.

A Universal Simulated Power Inductor
A continuously variable 1mH-10H device, without magnetism

Ramkumar Ramaswamy, PhD

1 Motivation and Overview

Inductors are the bugbear of audio design. They are unwieldy, difficult to get in close tolerances or with high Q-values; if high-valued, they are generally nonlinear due to the nature of the solid core. Therefore when one is prototyping a passive crossover, one generally tries to fix the inductance values beforehand and then play around with the capacitors and resistors to get things right. Even when a crossover is simulated beforehand or an active filter is used to arrive at the best combination of characteristic function and crossover points for a given system, there is a distinct need to play with these values as a final step to perfect the design of a passive crossover. It is with this context in mind that I set out to explore the world of simulated inductors (*real* simulated ones; not software-simulated ones!) and to build a practical one that could serve this need as well as others that come to mind.

If you want to build a simulated *grounded* inductor that does not have a very high Q, life is simple - a single op-amp and a couple of components is all you need (see e.g. [2]). However, I had four clear requirements the inductor had to meet:

- It must float
- It must handle at least 1A RMS
- It must have a reasonably high Q (low equivalent series resistance)
- It must be easily tuneable

With these requirements, things suddenly become considerably more complex, and you will not find too many practical circuits published.

I started by looking at two well-documented textbook approaches for the design of high-Q simulated inductors from which many variants have evolved:

Approach 1: Use op amps wired as Generalized Impedance Converters (GICs).

Approach 2: Use Operational Transconductance Amplifier ICs (OTAs) wired as gyrators.

Circuits for these two approaches are available in standard textbooks and reference books and for brevity are not reproduced in this article. (See, e.g. [2] for a detailed discussion on the performance of GIC-based inductors, and [1] for coverage on OTA-based inductors.) In each approach, you will find designs for two types of inductors: floating and grounded. A *variable floating inductor* is significantly more complex to design, and so I decided to focus on that.

These traditional approaches have some drawbacks. Specifically,

Figure 1: The simulated floating inductor mounted in a 3U subrack.

Drawback of Approach 1: The floating inductor requires two GICs connected back-to-back. Each GIC uses one capacitor so you would need two capacitors. Further, there are component-matching requirements, so changing the value of the inductor would require require changing component values in each GIC in a matched fashion which is bothersome.

Drawback of Approach 2: The use of OTAs considerably simplifies the design and does away with many of the constraints of Approach 1. However, the issue here is that choices for commercial OTA ICs are few, and these are very low current devices. For example, the LM13700 is a popular OTA but its transconductance output delivers less than 1 mA.

Over the last several years, several new and hybrid designs that use op-amps as well as OTAs have been published in research papers, but most of them talk of results in software simulation. I had trouble finding a real, practical implementation that demonstrably met my four objectives. Therefore I decided to design my own practical inductor without deviating fundamentally from the classical, well-studied design approaches and see how they perform. I was not disappointed, hence this article!

I chose to adopt Approach 2 (OTA-centric topology) to make for easier tuning, but rather than using commercial OTA ICs, my idea was to use the venerable Howland Current Pump (HCP) [4] as a building block. A brief of this idea appears in the Ideas for Design column of Electronic Design magazine [5]. A HCP is a bidirectional current pump driven by a differential voltage input – that is, it is a differential transconductance amplifier – that uses a regular op-amp as its active element. The use of standard op-amps rather than OTAs is what allows for a high current capacity. Rather than use specialized power op-amps for my prototype, I gave myself maximum flexibility by designing around non-specialized op-amps with a standard current booster output stage (within the feedback loop). With the resulting design I now describe, I was able to build a floating inductor that fits on a compact 10x10 cm PCB and covers the range 1mH to 10H with a 1A RMS capacity. The photograph shows the working implementation. Two pushbuttons act as x10 and x100 inductance multipliers and a single potentiometer allows for continuous variation over a factor of 10. (In the prototype I used a 12-way rotary switch instead of a potentiometer to conveniently switch precise E12 values.)

2 The Design

A HCP-based topology may be used for floating as well as grounded inductors. A grounded inductor requires 2 HCPs and a floating one requires three. **Figure 2** shows the overall circuit diagram of a floating simulated inductor using three HCPs. U1C buffers the voltage across C3/C7 because the HCPs do not have infinite input impedance, and also facilitates inductance tuning.

A HCP is characterized by its transconductance *G*. That is, if the input voltages at the inverting and noninverting inputs of the HCP are V^- and V^+ respectively, the output current *I* is given by

$$I = G(V^+ - V^-)$$

Assume for a moment that H1, H2 and H3 represent perfect transconductance amplifiers. Therefore assume that U1C is absent and that the voltage across C3/C7 is directly connected to the inverting input of H1 (and the noninverting input of H3). Standard textbooks tell us [1] that this circuit simulates an inductor of value

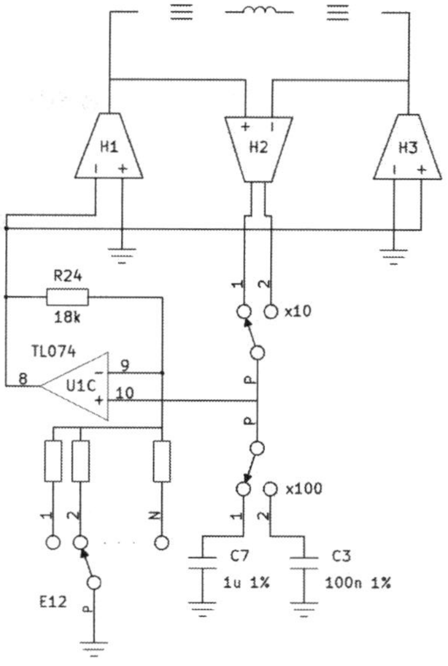

Figure 2: Overall schematic for the simulated power inductor

$$L = \frac{C}{G_1 G_2}$$

where G1 is the transconductance of H1 (which is required to be equal to that of H3) and G2 is that of H2.

In the implementation, H1 and H3 are power HCPs while H2 is a small signal HCP. The explosion for H1/H3 is shown in **Figure 3** and that for H2 is in **Figure 4**. It will shortly be clear why I show H2 as hav-

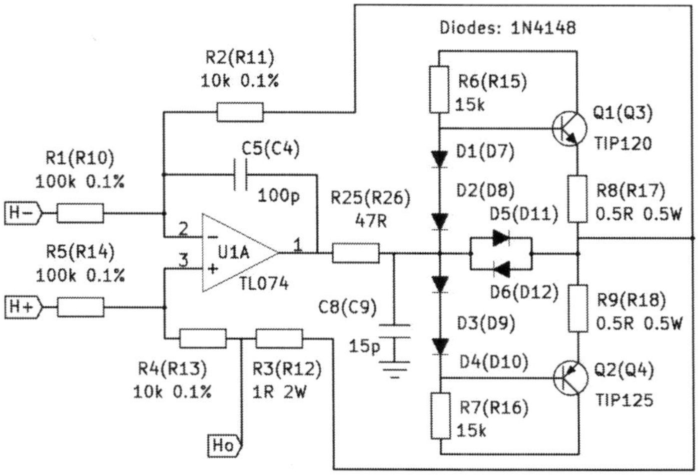

Figure 3: Explosion of H1 and H3, the power HCPs

ing two outputs. (In Figure 3, part numbers in parentheses refer to the corresponding part in the second power HCP.)

The currents through C3/C7 are extremely small – a few microamperes – and therefore we need U1C to buffer the capacitor voltage. U1C also allows us to continuously vary the inductor value; by increasing the gain of this buffer stage, we amplify the voltage across the capacitor. Since the capacitor is fed with a current from the constant-current source H2, a gain of *A* makes the capacitor (and hence the simulated inductor) look *A* times *smaller* than it actually is.

There are three independent "levers" we use to actually vary the inductance value

(a) The gain of U1C

(b) The transconductance of H2

(c) The value of the capacitor (C3/C7)

The reason I had to use capacitor switching rather than rely on (a) and (b) alone is because there is only a limited range within which I found I could vary these two levers in order to achieve a high inductance range while maintaining performance.

The power Howland stage

In Figure 3, H- and H+ are the inverting and noninverting inputs of the HCP, and Ho is the output. The paraphernalia surrounding the Darlingtons Q1 and Q2 is a standard current booster with current limiting incorporated around D5 and D6, so the output of the overall "power op-amp" is just the junction of R8 and R9. Bob Pease [4] recommends a low-value (~10pF) across the feedback resistor to the inverting pin (R2 of Figure 3 and R20 of Figure 4), but my tests indicated that a slightly better, though unusual approach is the use of C8 for the power HCP which I shall describe later. With the values shown, the value of the transconductance G1 is 100mA/V. Current limiting is set to 1A RMS by D5-D6-R8-R9.

The resistance ratios along the inverting and noninverting arms of the HCP must be equal. That is, in Figure 3 R2/R1 must equal (R3+R4)/R5 and in Figure 4 R20/ (R19a+R19b) must equal (R21+R22)/R23. The use of 0.1% resistors is recommended. Alternatively, the ratios can be equalized using a trimmer along any one arm. The 0.1%-resistor approach is usually sufficient, and easier to implement.

Design boundaries and tradeoffs

The capabilities of the design are governed by the following relationships which can easily be derived:

$$V_b = \frac{I_L}{G1} = 10\,I_L$$

where Vb is the voltage at the output of U1C and IL is the inductor current. These tell us that if we are content to work with a maximum inductor current I_L that does not exceed 1A RMS, then the buffer output voltage (and hence the voltage across C3/C7) will not exceed 10V RMS/14.1V peak. Further, looking at Figure 3, we see that I_L flows through R3 and therefore the output of the power op-amp exceeds the voltage at Ho by I_LR3 = 1VRMS. Therefore, as most standard op-amps are limited to an absolute maximum of ±18V supplies, we conclude that we can achieve a 1A current limit if G1 is *not*

much less than about 0.1A/V and R3 *is not much greater than* about 1 ohm, as is the case with the present design.
The voltage limit across the inductor is a corollary of the above limits: as long as the current remains within 1A, the inductor will operate in its linear range as long as the voltage across it does not exceed about 11VRMS with standard op-amps. These limits are satisfactory for most passive crossover prototyping needs where the normal test voltage is 2.828VRMS. (Caveat: the inductor voltage in a series L-C circuit at resonance can be much higher than the applied voltage.)
Note that increasing the inductance involves *reducing* the product G1G2. With G1=0.1, the formula says we can achieve an inductance of 10H if G2=1μA/V and C=1μF. This is a delicate balance: reducing G2 further will increase the susceptibility to noise and capacitor leakage, and increasing C is difficult owing to the limited choice of compact 1% polypropylene capacitors above 1μF.

The small-signal Howland stage

Figure 4 shows the explosion of H2. In this, I allow the output to be taken from two outputs Ho1 and Ho2. When the output is taken from Ho2, it is technically called an Improved Howland Current Pump. This gives us a HCP with two switchable values of the transconductance G2. Letting K ≡ (R21+R22)/R23 = R20/R19, these transconductances are

$$G2(Ho1) = K\frac{1}{R21+R22}$$

$$G2(Ho2) = K\frac{1}{R21}$$

Something interesting happens if we want this switch to effect a 1:10 ratio in inductance, and if we decide we want G2(Ho1)=1 μA/V so that we can achieve 10H. Taking G2(Ho2)= 10xG2(Ho1), we can now ***uniquely*** compute R23!

R21 = (R21+R22)/10 ⇒ R22 = 9R21
R23 = (R21+R22)/K = 10R21/K = 10/G2(Ho2) = 1M

Therefore R5 *must* be fixed at 1Meg; all we need to do is to choose R3 and R4 from standard values such that they are in the ratio 1:9 and all other resistors fall into place!

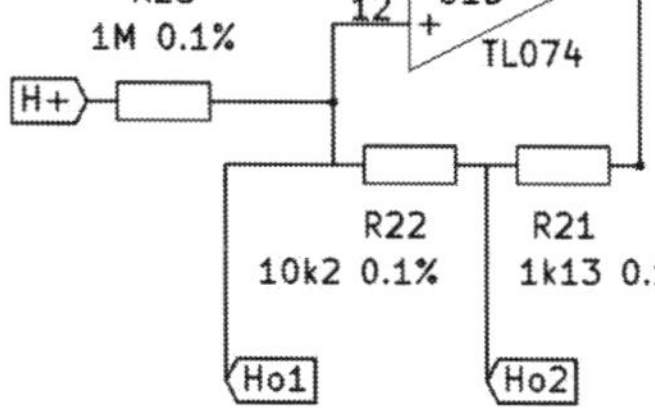

Figure 4: Explosion of H2

3 Performance Tests and Empirical Refinements

Though this is a floating inductor, the inductor input and output must still be "referenced" to ground, i.e., the surrounding circuitry must share its ground with the ground of this simulated inductor. Besides, it is not a good idea to connect one end of the floating inductor to ground directly as it can impair performance (it is better to use a *grounded* simulated inductor for this purpose). Therefore to reliably measure the inductance, it has to be put in series with something else. For example, you can put it in series with a 1mH inductor and then measure the total inductance. To get a better grip on the performance in a real crossover prototyping context, it is better to make measurements with a continuous AC signal rather than with time-constant measurement methods used by cheaper handheld LCR meters. Apart from the inductance value, we also want to measure the Q-factor.

An LC bridge is ideal for making these measurements, but sufficiently informative measurements can also be made with a simple series RLC network: you feed it with a sine wave and look for the resonance frequency *f* at which the voltage across the series LC combination is at a minimum. Since C is a known capacitor (I used 1% polypropylenes), the formula $f = 1/2\pi f\sqrt{(LC)}$ can then be used to calculate L. At resonance, the network becomes purely resistive (in principle). If *r* is the equivalent series resistance in the LC combination at resonance, the voltage across the LC combination will then just be Vr/(r+R) where V is the applied voltage. Since we know R and V, we can determine *r*. Then Q= $2\pi fL/r$. For this procedure, only the frequency counter needs to be accurate. The voltmeter need only have a high enough input impedance but for the purpose of finding L, need not be calibrated too well since all you are looking for is a *minimum*, not the actual voltage value. As for Q-factors, again we are not looking for the exact value, only an assurance that it is sufficiently high. **Table 1** summarizes the measured performance.

The Q-value and equivalent series resistance *r* of the inductor are stated as lower and upper limits respectively because the measured values include the effects of the equivalent series resistance of the test capacitor.

Compensation for capacitive loads

The power Howland stage was quite unstable even with mild capacitive loads such as a 1 meter oscilloscope probe. In the small-signal HCP of Figure 4, capacitor C6 stabilizes the circuit at high frequencies. However, the power HCP of Figure 3 may exhibit instability even if there is a capacitor across the corresponding feedback resistor R2 and that is because of the gradual ineffectiveness or breakdown of the feedback loop through the Darlington booster stage at MHz frequencies. A surer method is to adopt the standard capacitive load decoupling method C5-R25 *and* add the deliberate capacitive load C8. This effectively short circuits the op-amp output at those frequencies and prevents the booster stage from "participating" in the feedback to cause instability.

Front Panel Setting	Test Conditions	*f* (Hz)	Inductor Accuracy	Inductor Q @ *f*	Equivalent *r* @ f (Ω)
4.7mH	C = 6.8µF	890	Typically 1%	>20	<1.3
47mH	R = 1k	282		>40	<2.1
470mH	C = 23µF	23.2		>40	<3.3
4.7H	R = 1k	7.34		>20	<19

Table 1: AC performance summary

Output current offsets

At first you may think that since a current pump cannot be in equilibrium without *some* load, a simulated inductor that is built around current pumps must also need a load to be stable. However, if you look carefully at Figure 2, you will see that there is in fact a negative feedback loop that ideally keeps the inductor outputs at zero voltage. However, since our HCPs are not ideal transconductance amplifiers, in practice the feedback loop cannot ensure that the no-load output voltage is zero. In fact the no-load DC voltage can be several volts because of small offsets on the input side (partly contributed by the input offset of the op-amps) which causes a *current* output offset (as opposed to a voltage output offset). When the load resistance is high or infinite (open), even a small current offset can force the outputs to rise or fall several volts; if it approaches the supply rail the behavior can be embarrassing: fact I found that while operating it with a ±10V supply, no-load conditions caused *very* slow oscillations of a few Hz which stopped if I increased the supply to ±15V.

As long as things are not oscillating, the offset voltage *per se* is actually not a concern for passive crossover prototyping applications because the moment you connect a load to either terminal, the output voltage at both terminals fall to a harmless value. In my prototype, the no-load output voltage was 9V which fell to 0.2V when I connected a 360 ohm resistor and 20mV with a 36 ohm resistor, indicating that the equivalent output offset current was about 0.5mA. Admittedly there is scope for improvement in this parameter in future designs.

4 Construction

Component selection

The crucial components in this project are the precision 0.1% resistors and the precision 1% capacitors. You need not necessarily purchase precision components – if you have calibrated measuring equipment it is cheaper to obtain a batch of 1% metal film resistors of a reputed make and measure them till you get values within 0.1%. The same holds for the capacitors – however, it is best to stick with polypropylene for best results. It is difficult to find 1uF 1% polypropylenes; however, relatively cheap and accurate Dayton Audio caps (PMPC series) are available from Parts Express (www.parts-express.com) which is what I used for the main circuit as well as for performance testing.

Enclosure and PCB

A prototyping system using inductors must be a plug and play system that needs to accommodate different numbers of inductors at different points in time. Therefore I designed this on a 10x10 Eurocard-sized PCB with an edge connector for the power supply. This is fixed onto a standard 3U 10HP front panel which slides into a standard 3U subrack. The front panel can be made using commonly available waterslide decals, and finished with an acrylic spray.

5 Concluding Remarks

This prototype was primarily intended to prove the practicality of the concept of a universal, wide-range tuneable power inductor, specifically its floating *avatar*. However, as Jan Didden pointed out

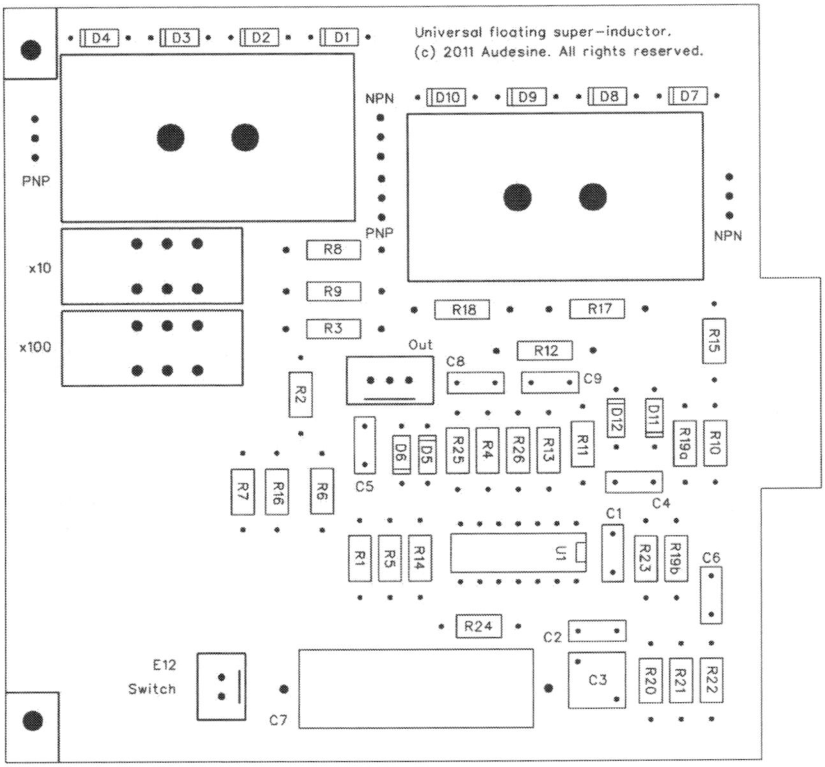

Figure 5: The PCB stuffing guide

over the course of many exceedingly helpful discussions, 1A may be sufficient for many crossover prototyping applications but certainly not all audio applications of interest. The next step for me is to therefore create an implementation that can handle an order of magnitude higher current. The changes to the design *per se* will not be significant; as I see it, the changes will be primarily the following:

- Adopting high voltage, high current power op-amps for H1 and H3
- Judiciously reducing the value of the resistor R3 in Figure 3 (which may potentially reduce the usable frequency range of the inductor)
- Redesigning for RF stability of the circuit in the event that a network like C5-C8-R25 cannot be used with the power op-amp

Last but not the least, while I have also designed and implemented a grounded version of this power inductor around HCPs, at the time of writing this article I have not fully convinced myself that a HCP-based *grounded* power inductor is necessarily a superior choice when compared to a traditional GIC-based grounded inductor.

Note: Half-kits, plans and design assistance for this project are available from the author who may be reached at rr@audesine.com

PARTS LIST

Capacitors (10% ceramic unless stated)

C1, C2	0.1uF	
C3	10n	1% pp/polystyrene
C4, C5	100p	
C6	10p	
C7	1uF	1% pp
C8, C9	15p	

Resistors (0.1% ¼ watt MFR unless stated)

R1, R10	100k	
R2, R11	10k	
R3, R12 1R	1R	1% 2W
R4, R13	10k	
R5, R14	100k	
R6, R7, R15, R16	15k	5%
R8, R9, R17, R18	0.5R	5% 0.5W
R19a, R23	1M	
R19b	15k	1%
R20	11.5k	
R21	1.13k	
R22	10.2k	
R24	18k	1%
R25, R26	47R	5%

(Note: if ordering mil-spec 0.1% resistors, order 1/8Watt)

Semiconductors

U1	TL074
Q1, Q3	TIP120
Q2, Q4	TIP125
D1-D12	1N4148

Switches/pots and connectors

E12 selector	1-pole 12-way OR pot
x10, x100	Alps SPUN191400
PCB Edge	Cinch 50-12A-30

Hardware etc.

Heatsinks (2)	Aavid 504222B00000G

References

[1] T. Deliyannis, Y. Sun and J. K. Fidler. *Continuous-time Active Filter Design*, CRC Press, 1999.

[2] L. P. Huelsman and P. E. Allen. *Introduction to the Theory and Design of Active Filters*. McGraw-Hill, 1980.

[3] National Semiconductor. *Op Amp Circuit Collection*. Application Note 31, Sep 2002.

[4] R. A. Pease. *A Comprehensive Study of the Howland Current Pump*. National Semiconductor Application Note 1515, January 29, 2008.

[5] R. Ramaswamy. *"Create A Simulated Floating Power Inductor Using Standard Op Amps"*, Electronic Design Magazine Vol 59 No.13, Oct 06, 2011. Penton Publishing, NY. Also available online at *http://electronicdesign.com/article/power/Create-A-Simulated-Floating-Power-Inductor-Using-Standard-Op-Amps.aspx*

A low noise laboratory-grade measurement preamplifier

Samuel Groner

This article presents the design of a low noise laboratory preamplifier, which is particularly aimed at noise measurement tasks. Its design features a single-ended JFET input stage, an active drain load for good power supply rejection and a novel input AC coupling network which attains, together with a DC servo, a second-order response. The achieved performance includes an equivalent input voltage noise density of 390 pV/rtHz (1 kHz) and 500 pV/rtHz (10 Hz), a frequency response of ±0.1 dB (10 Hz–100 kHz), a power supply rejection of at least 95 dB (10 Hz–100 kHz) and a quiescent current of just 17 mA. Detailed theoretical background on low noise design along with concise construction and application notes are included.

1 Introduction

Low noise is an important property of quality audio electronics; hiss that covers the details of a beautifully recorded reverb tail is not a welcome companion for the audiophile listener. Hence the ability to quantify noise is a prime need for a competent audio electronics laboratory. Available meters (typically either a true-RMS AC meter for wideband measurements or an FFT based spectrum analyser for spectral noise density measurements) however are often not sensitive enough to measure representative noise levels directly; therefore a preamplifier to boost the noise prior to the meter is needed. It is clear that such a preamplifier must be designed so that it does not introduce significant measurement errors. As will become obvious in the course of this article, this is anything else than trivial. In fact the design of a laboratory preamplifier for noise measurements represents a very interesting design challenge, and many of the applied techniques will be useful to advance more general audio circuit designs. In this article I'd like to not just present the theoretical analysis of a given design, but rather to show the evolutionary process which leads to an effective solution of the problem. While this makes the text lengthy, I hope that it aids the applicability of the presented approaches to other cases.

The following section compiles the detailed intended specifications for the low noise laboratory preamplifier. Later on, section 3 discusses some initial design considerations and in section 4 I will identify the various noise sources in a typical preamplifier topology and show how they contribute to the total equivalent input voltage noise. Sections 5 through 9 cover various performance improvements of the basic topology, e.g. with respect to power supply rejection. The actual design is dis-

cussed in section 10, and section 11 presents the measured performance. Section 12 and 13 complete this article with some construction and application notes, and a conclusion.
This article presumes a solid understanding of noise theory basics. Excellent reading are e.g. [1][2]; a shorter and more accessible text is [3].

2 Specifications

Every sound design process must start with a clear definition of the performance goal; only with well-defined demands it is possible to tackle the numerous trade-offs which will appear during circuit evolution. For this laboratory preamplifier design the primary goal is to accurately amplify noise, and the specifications shall ensure that no significant measurement error will occur for the intended range of application.
Before we can lay out detailed specifications we thus need to know how a preamplifier can introduce measurement errors; there are six primary error sources:

- The input voltage noise of the preamplifier adds to the noise of the source.
- The input current noise of the preamplifier flows into the source impedance; this causes a corresponding voltage noise term which adds to the noise of the source.
- Inaccurate preamplifier gain alters the measured value of the source noise.
- The frequency response of the preamplifier shapes the spectral distribution of the source noise and thus introduces a measurement error.
- Finite input impedance of the preamplifier loads the source and alters the effective voltage gain from source to meter.
- Finite output impedance of the preamplifier interacts with the input impedance of the used meter and alters the effective voltage gain from source to meter.

From this compilation we can derive that the specifications will need to focus on three areas in order to limit the measurement error: input voltage and current noise, gain and frequency response, and input and output impedance. The overall required performance depends essentially on the source characteristics, the performance of the used meter and the desired accuracy.
My main application field for this novel laboratory amplifier is the verification of low noise power supplies (which are e.g. needed to power AD and DA converters) with voltage noise density levels down to 1 nV/rtHz. The preamplifier should however also be applicable to the measurement of sources with output impedances up to 1 kΩ, e.g. to investigate the voltage noise behaviour of diodes or to measure general audio electronics. The frequency range of interest is 10 Hz to 100 kHz, and the total measurement error which the preamplifier introduces should be below 5%. To allow battery powering (which minimizes the effect from ground loops) quiescent supply current must be low, and the circuit should be insensitive to varying supply voltages.
The input referred voltage noise of typical meters is usually below 100 nV/rtHz in the frequency range of interest. A preamplifier voltage gain of 1000 (60 dB) thus makes the meter noise contribution negligible. Meter input impedance can be assumed to be high (well above 10 kΩ).
After an extensive feasibility study detailed specifications as shown in **table 1** were derived from the above outline.

Parameter	Conditions	Value/Limit
Input Voltage Noise Density	10 Hz–100 Hz 100 Hz–1 kHz 1 kHz–100 kHz	≤ 1.5 nV/rtHz ≤ 1.0 nV/rtHz ≤ 500 pV/rtHz
Input Current Noise Density	10 Hz–100 kHz	≤ 500 fA/rtHz
Voltage Gain	1 kHz	1000, ±0.5%
Frequency Response	10 Hz–100 kHz, R_S ≤ 1 kΩ	±0.1 dB
Input Impedance	1 kHz	≥ 100 kΩ
Output Impedance	unbalanced output balanced output	50 Ω, ±5% 100 Ω, ±5%
Output Voltage Swing	R_L ≥ 1 kΩ	≥ ±10 V
Power Bandwidth	V_O = 5 V_{PP}	≥ 100 kHz
Supply Voltage		±15 V to ±20 V
Quiescent Supply Current		≤ 20 mA

Table 1: Specifications for the low noise preamplifier.

It is easily seen that with the shown input voltage noise density specifications the intended 5% error budget is by far exceeded when measuring a 1 nV/rtHz source; the voltage noise density of the preamplifier would need to be below 320 pV/rtHz in order to meet the error budget. This however is, particularly at the given power consumption level and at 10 Hz, a very difficult goal. To greatly reduce the impact of preamplifier voltage noise it is however possible to subtract—in the usual root-mean-square fashion—the input voltage noise of the preamplifier from the initial measurement. If the voltage noise of the device under test approaches (or possibly even falls below) the contribution of the preamplifier, the subtraction process becomes very sensitive to minor measurement errors of the preamplifier noise level. For easy application of the preamplifier it is thus necessary that its input voltage noise is very stable, e.g. with time and supply voltage.
You may wonder why the input impedance is just stated at 1 kHz; this is sufficient because the frequency response, which must be met at the highest source impedance of 1 kΩ, implicitly constraints the input impedance variation with frequency to a sufficiently low value.

There is a large repository of prior art low noise preamplifiers [4–12] available. However, none of these designs fitted all specifications as detailed in table 1; in particular, no implementation could simultaneously guarantee the required gain accuracy and frequency response along with the desired voltage noise density levels and insensitivity to power supply voltage variations. Hence a new preamplifier design was envisaged; let's start with some initial design considerations in the next section.

3 Initial Design Considerations

A very fundamental design decision concerns the choice of transistor type for the preamplifier input stage. As MOSFET parts tend to have high excess noise corner frequencies, the available choices reduce to bipolar transistors (BJTs) and JFETs. BJTs may have very low excess voltage noise corner frequencies[1], which also can quite easily offer low wideband voltage noise levels. However, due to their considerable base current, they have inherently high current noise. Using a fundamental noise model of the bipolar transistor [1], it can be shown that a perfect transistor with zero base resistance must be run at least at 865 μA drain current to achieve a 500 pV/rtHz input voltage noise density level. Also easily calculated is that, for the desired 500 fA/rtHz input current noise density level, the base current of the transistor would simultaneously need to be below 780 nA. This implies a h_{FE} of over 1100; currently no commercially available discrete bipolar transistor does guarantee such a figure.
So JFETs are the way to go. For a given bias current level, their voltage noise is usually appreciably higher than for bipolar transistors; however, properly implemented, the current noise is several orders of magnitude lower. This allows us to parallel several transistors, thus lowering the voltage noise without exceeding the intended current noise specification.
Another major design consideration is the choice of input stage topology. In particular the decision between a single-ended architecture or a differential pair is fundamental. In short, the differential pair is superior in almost every respect (of particular importance for our application are its inherently low offset and good power supply rejection)—except that the presence of an additional transistor increases the voltage noise level by 3 dB compared to a single-ended implementation. As voltage noise performance is crucial for this preamplifier design, we relinquish the beautiful properties of the differential pair and opt for a single-ended front-end. With this decision we're ready to look more closely at the amplifier topology and the various noise sources of the preamplifier in the next section.

4 Basic Preamplifier Topology and its Noise Sources

Figure 1 shows a typical basic two-stage amplifier topology suitable for a low-noise preamplifier; the single-ended JFET input stage is formed by Q1 and drain load resistor R2. Operational amplifier U1 constitutes the second amplifier stage, and C2 the corresponding Miller compensation capacitor. V1 provides correct bias for Q1 while R3 and R4 establish global feedback from the output back to the source of Q1 (which serves as negative amplifier input). C1 and R1 AC couple the input to remove any DC content from the source. Also C3 limits amplifier gain to unity at DC in order to keep output offset reasonably low.

[1] I've been able to verify as low as 500 mHz for the Analog Devices MAT-12 transistor running at 1.6 mA collector current.

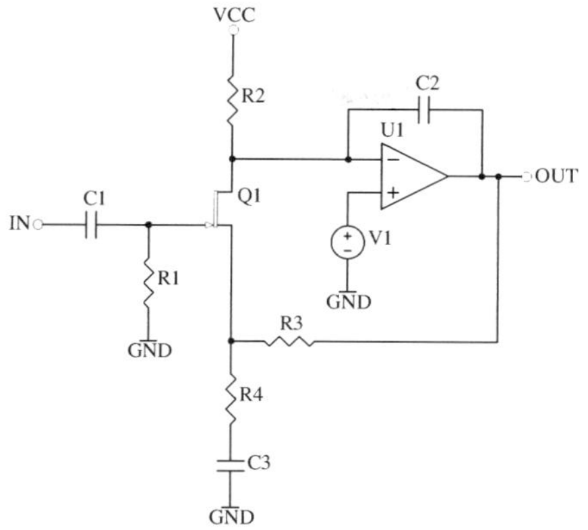

Figure 1: Basic preamplifier topology with single-ended input stage.

For an optimum design it is necessary to understand the contribution of the various noise sources to the total input referred noise of the preamplifier. In **figure 2** the equivalent voltage and current noise generators are shown for the basic two-stage amplifier topology introduced above. We can identify three groups of noise contributors; first of all, the resistors (R1 through R4) show thermal noise. Second, the active devices (Q1, U1 and V1) contribute with their equivalent input voltage and current noise generators.[2] Last but not least, noise injected from the positive power supply (VCC) affects performance.[3]

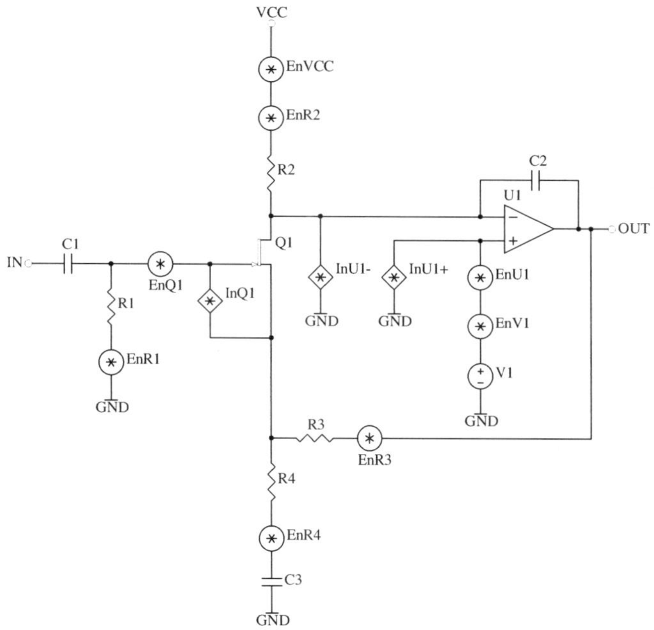

Figure 2: Noise sources in the basic preamplifier topology.

Reflecting these various noise sources back to the input to calculate their equivalent input voltage noise contribution is basically straightforward, but the mathematical expressions quickly become involved if no simplifications are done. As sufficient literature is available on the detailed analysis steps [1][11][13] I'm just giving the results of the analysis, along with the all-important interpretation. The simplifying assumptions (which all are valid for a typical implementation) are as follows:

- The impedance of C1 is much smaller than R1 at the frequencies of interest.
- The source impedance of the preamplifier is much smaller than R1.

[2] The voltage reference V1 is, for simplicity, treated as active device even though there are purely passive implementation possibilities.

[3] For the time being we will presume that the power supply rejection of the operational amplifier U1 is high enough, such it does not present a significant path for power supply noise to affect preamplifier performance.

- The impedance of C3 is much smaller than R4 at the frequencies of interest.
- R4 is much smaller than R3.

Also the discussion will focus on the fundamental white noise sources; the handling of low frequency excess noise is treated later.

Let's start with the noise contribution of R1:

$$e_{n|R_1} \approx \frac{E_{nR_1}}{2\pi f \cdot R_1 C_1} \tag{1}$$

$$\approx \frac{\sqrt{4k_B T \cdot R_1}}{2\pi f \cdot R_1 C_1} \tag{2}$$

Equation 1 shows that this noise contribution increases with falling frequency at 20 dB/decade. This behaviour is easily understood by considering figure 2; the voltage noise generator of R1 (E_{nR1}) is low pass filtered by R1 and C1 (recall that the source impedance is low compared to R1, i.e. the end of C1 facing the preamplifier input is effectively connected to ground) before it appears at the gate of Q1. Equation 2 includes the formula to calculate thermal voltage noise with Boltzmann's constant k (1.38×10^{-23} J/K) and absolute temperature T in Kelvin. It also reveals that the noise contribution of R1 *decreases* for higher resistance values. While the thermal voltage noise increases by 3 dB for a doubled resistance value, the attenuation from the low pass action of the AC coupling network is boosted by 6 dB. It results in an overall 3 dB noise reduction. From both the mathematical expressions and the interpretation it is also clear that increasing the capacitance value proportionally reduces noise.

The drain load resistor R2 is an interesting case as well:

$$e_{n|R_2} \approx \frac{E_{nR_2}}{R_2} \cdot \left(\frac{1}{g_{m|Q_1}} + R_4 \right) \tag{3}$$

$$\approx \frac{\sqrt{4k_B T \cdot R_2}}{R_2} \cdot \left(\frac{1}{g_{m|Q_1}} + R_4 \right) \tag{4}$$

Equations 3 and 4 can intuitively be read as follows: The negative input of U1 provides a virtual earth potential, which sums the drain current of Q1 with the currents flowing through R2 and C2; R2 is thus effectively connected to a low impedance current node, and its noise contribution is in the form of a current. The magnitude of this current is simply the thermal voltage noise generator of R2 divided by the resistance of R2. This current noise is referred back to the input by division by the net input stage transconductance (i.e. voltage-to-current gain); the net transconductance of the input stage is given by the transconductance of Q1, which is reduced by the source degeneration provided by R4. To minimise the noise contribution of R2 there are hence two options: increasing its value, or increasing the transconductance of the input stage.

Last but not least the feedback network contributes thermal noise:

$$e_{n|R_3,R_4} \approx E_{nR_4} \tag{5}$$

$$\approx \sqrt{4k_B T \cdot R_4} \tag{6}$$

As R3 is much larger than R4, their parallel combination (which forms the equivalent resistance for the total thermal noise contributor) is mostly dominated by R4, and R3 can be neglected. The thermal voltage noise of R3 then appears directly in series with the input. It is thus clear that a lower resistor value (with R4 scaled proportionally to keep preamplifier gain constant) will reduce noise. From equations 3 and 4 it is seen that this is also advantageous to reduce the contribution of the drain load resistor, as the net input stage transconductance is increased; other noise sources will benefit from this change too, as will emerge later in this section.

It is worth repeating that from the three discussed resistor noise contributions only one reduces with smaller resistance values; the other two require *high* values for low noise! Surely the sometimes indiscriminately applied generalisation that for low noise low resistance values must be used is not valid.

Let's look further and check the noise contribution of the input transistor Q1:

$$e_{n|Q_{1e}} \approx E_{nQ_1} \tag{7}$$

$$\approx \sqrt{\frac{8}{3}kT \cdot \frac{1}{g_{m|Q_1}}} \tag{8}$$

$$e_{n|Q_{1i}} \approx \frac{I_{nQ_1}}{2\pi f \cdot C_1} \tag{9}$$

$$\approx \frac{\sqrt{2q \cdot I_{G|Q_1}}}{2\pi f \cdot C_1} \tag{10}$$

As is clear from figure 2 and as equation 7 shows, the equivalent input voltage noise term of the JFET appears directly at the preamplifier input. Equation 8 extends this with the inclusion of the van der Ziel estimate for the JFET equivalent thermal channel resistance [14]. It shows that for low voltage noise above the excess noise region the JFET should have high transconductance; this again implies that the transistor should be run at high drain current, as transconductance increases approximately with the square-root of drain current [15].

A further contribution to the equivalent input voltage noise is given by the current noise of Q1, which flows through the AC coupling capacitor where it causes an equivalent voltage noise term; this is illustrated by equation 9. In figure 2 it is seen that the current noise of the input transistor necessarily flows through the feedback network as well. As the impedance of the latter is very low, this contribution is entirely negligible though.

At low frequencies (where the impedance of the AC coupling capacitor C1 is high and thus the current noise of Q1 becomes relevant), the current noise of a JFET can be estimated directly from the gate current I_G and corresponding shot noise [1]; equation 10 includes this estimate, where q is the electronic charge of 1.60×10^{-19} C. From this it is seen that the gate current of the input transistor should be minimised (the means for this are detailed in section 5), and a sufficiently large value for C1 is necessary.

The contribution of the operational amplifier is more complicated to analyse:

$$e_{n|U_{1e}} \approx E_{nU_1} \cdot \left(\frac{1}{R_2} + g_{o|Q_1}\right) \cdot \left(\frac{1}{g_{m|Q_1}} + R_4\right) \tag{11}$$

$$e_{n|U_{1i-}} \approx I_{nU_1-} \cdot \left(\frac{1}{g_{m|Q_1}} + R_4\right) \tag{12}$$

$$e_{n|U_{1i+}} \approx I_{nU_1+} \cdot Z_{V_1} \cdot \left(\frac{1}{R_2} + g_{o|Q_1}\right) \cdot \left(\frac{1}{g_{m|Q_1}} + R_4\right) \tag{13}$$

Equation 11 can be understood as follows: Through the action of feedback, the two inputs of the operational amplifier are kept at the same voltage. This means that the equivalent input voltage noise generator E_{nU1} modulates the voltage at the drain of Q1. The resistance at this node, given by the parallel combination of R2 and Q1 output resistance, converts this voltage to a current which is summed with the other currents flowing into this node. The resulting noise contribution is referred back to the input once more by division by the net input stage transconductance. For mathematical simplicity, R2 and Q1 output resistance are expressed in their corresponding reciprocal forms, i.e. as conductances.

The input current noise from the inverting operational amplifier input terminal again sums directly with the other currents present at this node, and is referred to the input by division by the net input stage transconductance. This is highlighted by equation 12. In contrast, equation 13 shows that the input current noise from the non-inverting operational amplifier input terminal flows through the impedance Z_{V1} of V1. The resulting voltage noise term appears effectively in series with the input voltage noise generator of U1, and is thus referred back to the input as described above. From equation 11–13 it is seen that—besides selecting a part with low voltage and current noise to begin with—operational amplifier noise contribution is very effectively reduced by increasing net input stage transconductance and the resistance present at the current summing node (i.e. the non-inverting input of the operational amplifier).

The same line of thought applies to the voltage noise generator of V1:

$$e_{n|V_1} \approx E_{nV_1} \cdot \left(\frac{1}{R_2} + g_{o|Q_1}\right) \cdot \left(\frac{1}{g_{m|Q_1}} + R_4\right) \tag{14}$$

The analysis is completed by considering the noise induced from the positive power supply:

$$e_{n|VCC} \approx \frac{E_{nVCC}}{R_2} \cdot \left(\frac{1}{g_{m|Q_1}} + R_4\right) \tag{15}$$

We now see that the equivalent input noise of a preamplifier is a complex function of several noise sources. There are many interactions between the various design parameters, which complicates the optimum choice of part types and bias conditions. Clearly it is not realistic to find a well-defined, simple global optimum for the total preamplifier equivalent input voltage noise which is realisable with a unique and practical set of design parameter choices.

It can be said however that the most intrinsic (and hence most difficult to reduce) noise contribution is that from the equivalent input voltage noise of Q1. Thus it makes sense to accept this as the dominant (i.e. primary) noise source, and make sure that all other noise contributors remain at considerably lower level (i.e. are secondary). Usually I try to design a circuit such that all secondary noise sources are at least 5 times, preferably 10 times, lower than the primary noise source. As there is a rather large number of secondary noise sources, this is less conservative than might be expected at first glance. Also such a design strategy makes the overall design robust against outliers in the secondary noise sources.

With this in mind, the following sections will show how the architecture from figure 1 can be evolved—by parts and bias condition choice as well as topological changes—into a high-performance preamplifier meeting the specifications as given in section 2. I'll start with the discussion of the input stage design.

5 Input Stage Design

Upon reviewing equations 3–15 it is clear that the amplifier input stage, consisting of JFET Q1 and its drain load R2, has an exquisitely important role for overall noise performance. In particular we want:

- Q1 to have the lowest possible voltage and current noise.
- Q1 to have the highest possible transconductance to reduce various secondary noise contributors.
- R2 to be as large as possible in order to minimise its own and several other secondary noise contributions.
- the output conductance of Q1 to be considerably lower than the reciprocal of R2, again to minimise multiple secondary noise sources.

To understand how this may be achieved, we need to look more closely at the behaviour of available JFETs and the interaction with the drain load resistor.

As noted above, JFET transconductance increases approximately with the square-root of drain current. Equation 8 thus indicates that voltage noise should drop with one-fourth the power of drain current [16]. Actual measurements however show considerably lower improvement with drain current; **figure 3** illustrates this. The measurement data (at 1 kHz) is taken from the 2SK170 data sheet [17], and the calculated characteristic is made to fit the noise performance at the lowest drain current of 200 μA. At 10 mA drain current, the measured voltage noise density value is about 1.6 times higher than the value predicted by the one-fourth power law. There are several reasons for this behaviour:

- The transconductance towards high drain currents is lower than predicted by the fundamental square-law [18]. At I_{DSS}, the deviation may be as much as 25% or more. Lower transconductance results in higher voltage noise, as discussed above.
- The thermal dissipation of the JFET, and thus its operating temperature, increases with drain current. As equation 8 shows, this will also increase observed voltage noise.

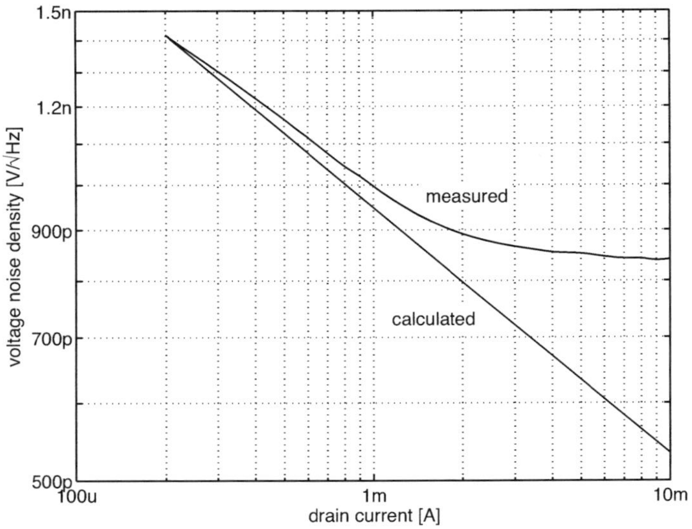

Figure 3: Measured vs. calculated equivalent input voltage noise density for a typical JFET.

- There are small bulk resistances associated in series with the gate and source [19]; these resistances are presumably mostly independent from drain current and their thermal noise contribution not included in equation 8.[4]

The magnitude of these effects and thus the deviation from the theoretically predicted noise performance varies appreciably for different part types. The measurement data from figure 3 can be considered typical for contemporary low voltage noise JFETs, but vastly more significant deviations were reported in the literature [1]. In fact, for certain parts and operating conditions even increased voltage noise towards high drain currents is observed [15].

Besides thermal channel noise, a real JFET will also exhibit low frequency excess voltage noise. The magnitude, exact spectral shape and dependence on drain current is not easily predicted analytically, but should be experimentally determined for a specific part. Basically excess noise decreases with increasing drain current, just as thermal channel noise does. However the gain for higher quiescent current is typically even lower [20].

At this point it is important to realise that, in order to increase input transistor drain current, it is necessary to proportionally reduce the drain load resistor (unless further circuit changes, such as increasing the power supply voltage, are carried out). This however is conflicting with the requirements given at the beginning of this section where we stated that R2 should be chosen as large as possible, because many secondary noise sources decrease with higher values of R2 (see equations 3, 4, 11 and 13–15). The very same secondary noise sources however also reduce proportionally with the net

[4] The presence of bulk resistance at the source also explains, at least partially, the deviation from the nominal square-root transconductance characteristics.

input stage transconductance. As the later increases at higher drain currents, there are two opposing effects. At first it is thus not clear if increasing the drain current of the input transistor amplifies or attenuates, or possibly even does not alter, the influence of secondary noise sources.
The dust clears quickly as we—once more—recall that the transconductance of the JFET nominally increases with the square-root of drain current. The effective increase in net input stage transconductance is even lower than this for two reasons: first, actual JFETs show lower transconductance gain as discussed above; second, the feedback network degenerates the common-source configuration of Q1 (as shown in equations 3, 4, 11 and 13–15) and thus further limits the net transconductance increase. Now let's assume that the reciprocal of R2 is much larger than the output conductance of Q1; the various secondary noise sources described in equations 11 and 13–15 thus increase proportionally with the reciprocal of R2—if the value of R2 is halved, the contribution of these noise sources to the equivalent input noise doubles. It is now clear that the increased net input stage transconductance at higher drain currents cannot make up for the effect of reduced drain load resistance; the various secondary noise sources increase by more than 3 dB for each doubling of drain current.
So the benefit of lower JFET voltage noise at higher drain currents is counteracted by the increased secondary noise sources; this means that there exists an optimum drain current, above which the total input referred voltage noise of the preamplifier deteriorates because the secondary noise sources dominate overall performance.[5] **Figure 4** presents a calculated example which includes the noise contributions of JFET voltage noise, drain load resistor, feedback network and operational am-

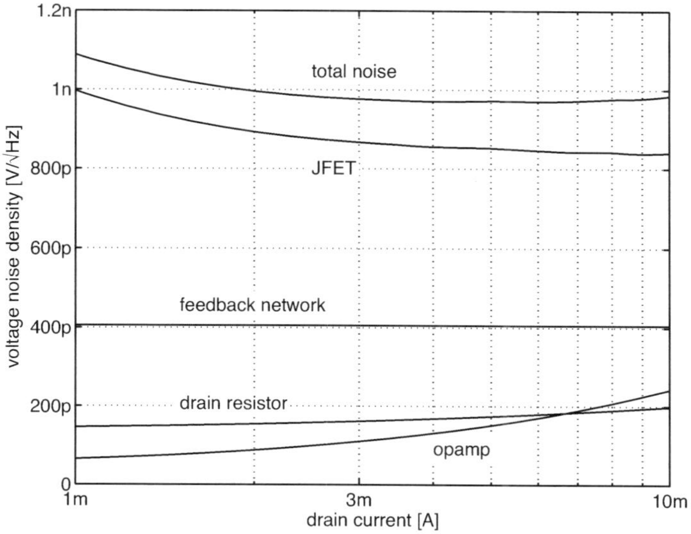

Figure 4: Total input referred voltage noise density vs. drain current.

[5] By putting sufficient efforts into minimising the various secondary noise sources it is however possible that the optimum drain current lies above I_{DSS} and thus is of no practical relevance.

plifier voltage and current noise. For the JFET, voltage noise (at 1 kHz) and transconductance measurement data is taken again from [17]; the drain load resistor is set for 5 V DC drop (e.g. to 5 kΩ for 1 mA drain current); for the operational amplifier, input referred noise sources of 3.5 nV/rtHz and 400 fA/rtHz (the typical figures at 1 kHz for a NE5534A) are assumed. As is seen from figure 4, this configuration shows the predicted optimum at a drain current of about 5 mA. At the used frequency of 1 kHz the optimum is not particularly distinct, however at lower frequencies where excess noise comes into play it will typically be both more pronounced and occur at lower drain current. Measurement results presented later will confirm this.

This is a very essential result, and summarises the main difficulty encountered with JFET input stages. It follows that reducing voltage noise by increasing drain current is an approach with limited benefit towards high drain current values; this is particularly valid if low-frequency noise performance and power consumption is a consideration. As a rough guideline, biasing JFETs above one-half I_{DSS} is not advised. In contrast, the transconductance of bipolar transistors increases (near-perfectly) in proportion to collector current, which makes the influence of second stage noise sources far less significant—even at high collector currents.

Does paralleling JFETs bring us any more benefit than increasing the drain current of a single transistor? When we double the number of JFETs (and also double the total drain current by halving the drain load resistor value), the total transistor voltage noise reduces by a factor of 1.41 (or 3 dB). The transconductance however doubles—and thus cancels the penalty from the reduced drain load resistance value! Only the degenerative action of the feedback network causes a slight overall input stage transconductance limitation. This is a pleasant result; the contribution of the secondary noise sources is nearly independent from the number of paralleled JFETs. This means that, instead of using a low number of JFETs run each at high drain current, it is much more beneficial to select a larger number of transistors each biased at lower drain current. The number of JFETs is only limited by the tolerable input capacitance and current noise. This will be discussed just below and in section 9.

Deriving the detailed considerations to achieve low voltage noise has been rather laborious; fortunately current noise of JFETs is easier to handle. As equation 10 predicts, it depends solely on the magnitude of the gate current, and for the application at hand this is a sufficiently accurate approximation. Hence all we need to do is minimise gate current. For this it is helpful to appreciate that [15]:

- gate current is to a large extent independent from drain current;
- gate current typically doubles with every 10° K rise in temperature;
- gate current often increases drastically with drain-source voltage above a certain threshold, due to impact ionisation;
- maximum gate current is usually very conservatively specified by the manufacturer, due to the difficulties of actually measuring it. Typical values are well into the lower pA range, and there is no drastic performance difference amongst typical contemporary low voltage noise JFETs.

The most important means to keep current noise low is thus to operate the transistor at low drain-source voltage (V_{DS}). This minimises both operating temperature and the effect from impact ionisation. Because a low drain-source voltage allows highest values for the drain load resistor for a given

power supply voltage, and because it reduces the operating temperature of the JFET, this bias condition is also beneficial for low voltage noise.
A lower limit for the drain-source voltage is given by the output conductance of the JFET, which increases quickly as V_{DS} approaches the magnitude of $V_{GS(off)}$ (the gate-source cut-off voltage, sometimes also referred to as pinch-off voltage). So far we have assumed that the effect from output conductance is negligible compared to the drain load resistor. This is, in most practical cases, valid if V_{DS} is set to about twice the maximum specified $V_{GS(off)}$ magnitude.
Let me briefly summarise the most important findings of this section:

- For low voltage noise, it is beneficial to choose the highest number of paralleled JFETs which is still compatible with input capacitance and current noise constraints;
- There is an optimum drain current value which minimises total voltage noise. For typical configurations, the value is below I_{DSS} and a function of frequency. Towards low frequencies the optimum drain current value usually decreases;
- Biasing the input transistor(s) such that V_{DS} equals about twice the maximum specified $V_{GS(off)}$ magnitude is beneficial for both low voltage and current noise while keeping output conductance sufficiently high.

Armed with this insight, we're ready to explore the next challenge—making the preamplifier performance independent from noise on the power supply.

6 Power Supply Rejection

Equation 15 gives the input referred contribution of noise from the positive power supply; if E_{nVCC} is replaced with the constant one, the reciprocal of the resulting formula gives the basic positive power supply rejection of the amplifier topology from figure 1. Let's put in some numbers: if we take the circuit configuration used for figure 4 at the optimum drain current of 5 mA, then the drain resistor becomes 1 kΩ and the transconductance of the 2SK170 is about 30 mS. Together with the value for R4 (10 Ω as given in section 5) the power supply rejection is calculated—as a meagre 27 dB! Note that this figure is independent from frequency. Anyone used to read operational amplifier data sheets will agree with me that this state of affairs is disappointing; even the cheapest general-purpose monolithic amplifier will do an order of magnitude better at 10 kHz, and offer 100 dB or more at low frequencies. The poor power supply rejection is a direct consequence from the use of a single-ended input stage; if a differential pair is employed (as in standard operational amplifiers) power supply noise becomes a common-mode signal which is rejected by the differential nature of the input stage.
What can be done to improve the supply rejection of the single-ended input stage? Equation 15 suggests making the drain load resistor and net input stage transconductance higher; from section 5 we know that this is a contradicting demand, and that the use of paralleled JFETs, each run at modest drain current, will get most out of the situation. For realistic implementations, we can expect at best an order of magnitude improvement in power supply rejection from such changes. This means that further gain must be achieved by topological changes. Before we look into this, we should first define the actual power supply rejection requirements.

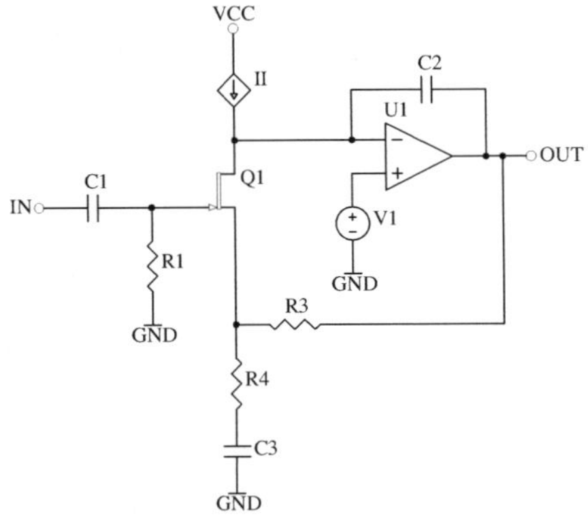

Figure 5: Preamplifier topology with active drain load.

In many cases, the laboratory preamplifier will be battery powered to reduce the effects from ground loops. Usually batteries are considered to show very low noise; this however is only true for certain types, when sufficiently charged and when operated with low current draw. Also it is convenient if the preamplifier can be powered from a standard laboratory power supply, e.g. for extended measurement periods or less critical work. Hence the power supply rejection of the preamplifier should be good enough to support this. I have measured the voltage noise of several linear laboratory power supplies; typically they show less than 10 µV/rtHz at 10 Hz, with rapidly decreasing voltage noise density towards higher frequencies. If the preamplifier has a power supply rejection of 90 dB at 10 Hz, this figure translates to an input referred contribution of 316 pV/rtHz. This is nearly one-fifth of the specified total preamplifier noise of 1.5 nV/rtHz at 10Hz and thus in agreement with the design strategy outlined at the end of section 4.

While prior art low noise preamplifiers have employed various means (e.g. the use of a low noise voltage regulator for the positive supply [4]) to achieve high power supply rejection, there is one straightforward approach which so far has not seen broad application ([30] being a rare exception): as increasing the value of the drain load resistor leads to better power supply rejection, replacing it with a constant current source will eliminate this power supply injection path. Contemplating **figure 5** makes this easier to understand; if the positive power supply is modulated, the current flowing through the constant current source is—by definition—constant. Hence no supply related current can flow into the current summing node (i.e. the inverting input of U1). Mathematically speaking, the value of R2 is replaced with the output impedance of I1 in equation 15. Of course a practical current source implementation cannot achieve infinite output impedance, so power supply rejection will still be limited. However a three orders of magnitude improvement over the plain drain load resistor is feasible.

The increased impedance at the current summing node also helps reducing other noise sources; in equation 11, 13 and 14 the value of R2 is again replaced by the output impedance of I1. Of particular value is the reduction of operational amplifier voltage noise. For a decent constant current source implementation the output conductance of Q1 becomes the limiting factor in this respect.

The noise contribution of the current source is most easily modelled by a parallel current noise source I_{nI1}. Equation 3 and 4 are then replaced by:

$$e_{n|I_1} \approx I_{nI_1} \cdot \left(\frac{1}{g_{m|Q_1}} + R_4 \right) \qquad (16)$$

The constant current source implementation inevitable uses active devices, which all contribute to I_{nI1}. In [21] I have presented the analysis of noise sources in folded cascodes, which was shown to be applicable to a standard current source implementation using a bipolar pass transistor. Let's recall the basic design guidelines which lead to low noise:

- Maximise the value of the emitter resistor by the use of a voltage reference with large DC voltage;
- Use a voltage reference with low noise and low impedance;
- Choose a pass transistor with low current noise, i.e. high h_{FE} and low excess noise.

The first requisite is, under the given constraint of available power supply voltage, easy to fulfil. The choice of a suitable reference voltage element is more difficult; forward-biased diodes have low noise, yet that comes with low DC voltage and relatively high impedance. Zener diodes, as well as bandgap references, offer a wide range of high output voltages and low impedance, but suffer from much higher noise levels. In many cases the use of a RC low pass filtered Zener diode will be the most practical path. Finding a suitable transistor type which offers sufficiently low excess noise is difficult. This is particularly because the pass transistor runs at relatively high collector current—a condition which pronounces excess noise [1]. However it is possible to replace the single transistor with a Darlington configuration; that way, the current noise performance of the composite transistor is dominated by the input transistor which runs at low current.

Figure 6 illustrates the concept of such an implementation. You may wonder what the optimum choice of R2 (which sets the collector current of Q2) is; a detailed analysis is beyond the scope of this article; however a good approximation is given by setting the collector current of Q2 such that the noise figure of Q2 is optimised for the value of R1. Some (actually: very few) transistor data sheets provide convenient noise figure curves which make the optimum choice simple. In the absence of such detailed measurement data the following approximation [22] is applicable:

$$I_C \approx \frac{26\,\mathrm{mV} \cdot \sqrt{h_{FE|Q_2}}}{R_1} \tag{17}$$

It should be appreciated that the optimum collector current is actually a function of frequency. At low frequencies, where excess noise is a significant contributor, the optimum collector current will be lower than in the white noise region.

Besides the pass transistors, there is another source of low frequency excess noise; unlike the other resistors used in the preamplifier topology from figure 5, the emitter resistor of the constant current source is exposed to a large DC voltage drop. As resistor excess noise is proportional to DC voltage [1] this represents a potentially perilous condition. A typical quality thin film resistor specifies excess noise between 0.01 µV/V and 0.1 µV/V [23]. If we take the circuit conditions used before (current source emitter resistor 1 kΩ with 5 V DC across) this translates to an equivalent current noise of 10.4 pA/rtHz to 104 pA/rtHz at 10 Hz. With a JFET transconductance of 30 mS and a feedback shunt resistor of 10 Ω, equation 16 refers this

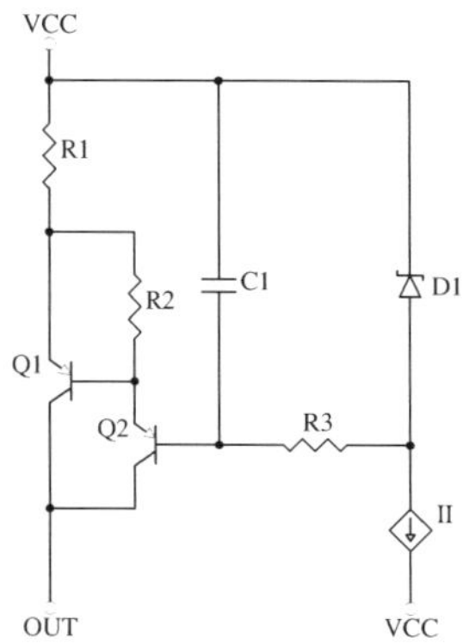

Figure 6: Constant current source topology suitable for low output current noise.

to the input as 452 pV/rtHz to 4.52 nV/rtHz. Clearly this noise contribution needs attention. Fortunately there is some relieve; the figures quoted above appear to be maximum values, as typical values are found to be at least an order of magnitude lower [24]. Still it is very advisable to look out for further improvement, as in the worst case the preamplifier specifications would be exceeded nearly three times. Spending money on precision resistors is not going to be promising as even the best conveniently available bulk metal foil and wirewound parts do not offer considerably lower guaranteed excess noise (see e.g. [25]). So what about lowering the DC voltage across the resistor? This is in contradiction to the design rules presented above, and thus not feasible. However, we can put several smaller resistors in series to synthesise the desired large resistance value; by doubling the number of resistors the DC voltage across each, and thus its excess noise, is halved. The noise contributions of the individual resistors add up in the usual root-mean-square manner, and thus a 3 dB excess noise reduction follows. The exact analysis is more complicated and depends on the resistor type used, as excess noise varies somewhat with resistor value. For most resistor types, excess noise reduces with lower resistance values. Using 10 resistors in series, an improvement in excess of 3 times is thus feasible, at low additional cost.

Using the presented constant current source topology and design guidelines, it is routinely possible to push the noise performance of an active drain load to a level where it is very close to the thermal noise contribution of the emitter resistor R2—in other words, nearly the same as that of a simple resistive drain load. Measurement results shown in section 11 will provide evidence of this claim.

The finite power supply rejection of the second stage (i.e. the operational amplifier U1) is usually only of minor concern; this is both because typical operational amplifiers show good power supply rejection within the frequency range of interest, and because power supply noise injected into the second stage is effectively suppressed by the gain of the input stage. As a simple precaution against the falling power supply rejection of operational amplifiers at higher frequencies, it may be useful to apply a modest amount of RC filtering at the amplifier supply pins.

Having cured the power supply rejection affliction of the basic topology from figure 1, and at the same time reduced the contribution of other secondary noise sources, it's time to contemplate some subtle issues associated with AC coupling.

7 AC Coupling

AC coupling at the input of the preamplifier is necessary to remove the potentially large DC signal of the source. Rendering the noise contribution of the AC coupling network negligible appears to be pretty simple; as noted before, equations 1, 2, 9 and 10 demonstrate that by choosing sufficiently large values for both resistance and capacitance (R1 and C1 in figure 1, 2 and 5) will provide arbitrary low noise levels from the AC coupling network. Unfortunately, there are two limitations which make the choice of appropriate component values more difficult:

- If the input of the preamplifier is connected to a source with high DC potential, large peak currents will flow through the AC coupling capacitor and subsequently through the input transistor. The magnitude of these currents is proportional to the AC coupling capacitor value and potentially very large. To reduce the risk of damage to the preamplifier, the AC coupling capacitor should be as small as possible;

- Leakage paths of the PCB material appear in parallel with the AC coupling resistor and limit the maximum resistance which is consistently achievable. Resistance values much above 100 MΩ typically require inconvenient precautions (e.g. guarding and PCB cleaning [26]).

So the seemingly simple choice of two passive components requires a bit more thought. First of all we need to know the rough order of magnitude of the input transistor current noise. If we assume a total gate current of 1 nA (this is a conservative figure which gives a safety margin against JFET outliers with a-typically high gate current and for operation at elevated temperatures), the nominator of equation 10 calculates as 17.9 fA/rtHz. Evaluation of the full equation 10 suggests that C1 should have a value of at least 1 µF in order to push the current noise contribution to less than 300 pV/rtHz at 10 Hz (one-fifth the input voltage noise specification of 1.5 nV/rtHz).

Using equation 2, it follows that the value for R1 needs to be at or above 47 MΩ such that the thermal noise contribution falls below 300 pV/rtHz at 10 Hz. This seems a reasonable set of values—the resistance is below the critical range, and the capacitance at least not very large. What is the cut-off frequency of this RC combination? It is a surprisingly low 3.39 mHz, with a corresponding time constant of nearly 5 minutes. This rises two problems; first of all, full settling of preamplifier output DC voltage after connection of a source with significant DC content may take several dozen minutes. Second, a source may show significant signal energy in the mHz region, e.g. due to short-term thermal effects. This signal content is not filtered by such an AC coupling network and may cause output overload of the preamplifier.

There is a nifty technique (which I'm not sure where it originates from, see e.g. [27]) which shifts the effective cutoff frequency of the RC network upwards without altering the component values (and thus the noise contribution). By connecting R1 not to ground but to a node which provides an amplified and inverted version of the signal present at the other end of R1 (i.e. the AC coupled input signal), a much lower resistance value is synthesised. **Figure 7** illustrates a possible implementation. The net resistance value of R1 calculates as follows:

$$\widehat{R_1} \approx \frac{R_1}{1 + \frac{R_6}{R_5} \cdot \left(1 + \frac{R_3}{R_4}\right)} \quad (18)$$

The signal gain of the inverting amplifier configuration of U2 will usually be kept to unity or lower, such that the output of U1 overloads earlier than that of U2. Even with this restriction, the cut-off frequency of the AC coupling network is easily shifted upwards by three orders of magnitude for a preamplifier with a gain of 60 dB, which fixes the two issues noted

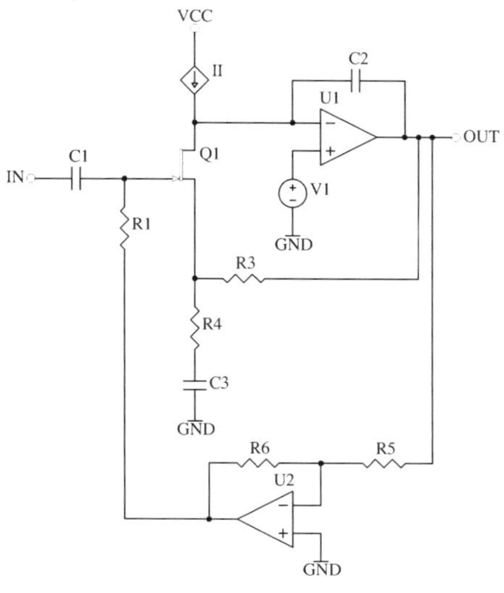

Figure 7: Topology with reduced AC coupling time constant.

above. In fact, frequency response specifications may force a lower cut-off frequency than achievable with this technique. Thus even higher values for R1 may be applicable, which further reduces its noise contribution. It should be noted that the noise contribution of U2 (and its feedback network) is not significant, as it is low pass filtered by R1 and C1 and generally speaking much lower than the thermal noise of R1 anyway.

The AC coupling capacitor C3, which limits amplifier gain to unity at DC, is inconvenient as well. As R4 is chosen as small as possible for best noise performance, C3 will need a large value which is only realisable with electrolytic capacitors. Their equivalent series resistance (ESR) is relatively high, varies from part to part, may be a strong function of temperature and can increase considerably with age. As the equivalent series resistance appears in series with R4, it will negatively affect the gain accuracy of the preamplifier.

An alternative technique to keep output offset low is the use of a DC servo. Basically there are three possible nodes where the servo could inject its correction signal; the gate, source or drain of the input transistor(s). The source is inconvenient as it is a node with very low impedance, and thus requires large currents to enforce a sufficient DC correction. The drain is neither suitable as the servo would then modulate the drain current of the JFET(s) to correct the DC offset—as we have seen, the drain current magnitude is crucial for good noise performance and thus needs to be well controlled. The gate however appears to be a decent choice; figure 8 reveals that all we need to do is relocate C3, because U2 happily agrees to also act as DC servo.

The time constants of C1-R1 and C3-R5 interact and convert the AC coupling to a second order high pass filter. This is of great benefit, as for a given cut-off frequency the second order filter provides both better pass band flatness and higher rejection of very low frequency signal content. I freely admit that so far I have been too lazy to calculate the required component ratio conditions to achieve a maximally flat (i.e. Butterworth) response; the values used in the actual preamplifier circuit presented later were derived by simulation.

As far as I know the topology shown in figure 8—with its significant advantages of high power supply rejection, low noise second order input AC coupling network and convenient DC servo integration—is shown here for the first time. It is however difficult to survey all prior art, and I'd appreciate reader comments on this.

After the significant topological evolution in this and the previous section there's not much missing in the puzzle of low noise preamplifier design; the next two (much shorter!) sections will mainly cover the handling of operational amplifier noise and bandwidth limitations.

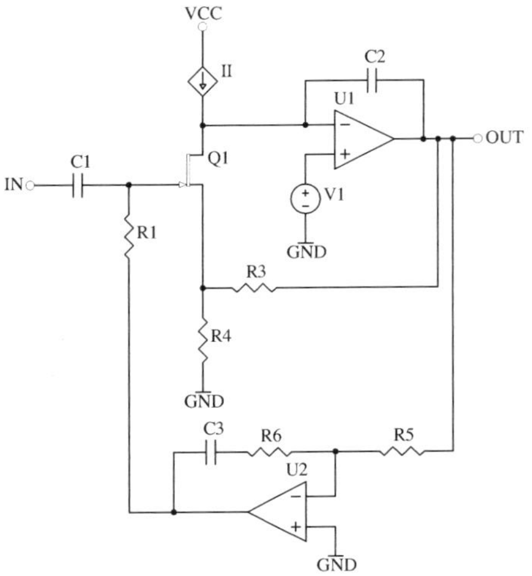

Figure 8: Preamplifier with DC servo.

8 Operational Amplifier and Voltage Reference

We have already seen that by a careful design of the input stage the noise contribution of the operational amplifier can be greatly reduced. Nonetheless it pays off to select an amplifier with inherently good noise behaviour. Typically there is a tradeoff associated with voltage and current noise; amplifiers excel either in one or the other (and in some cases, in neither...). Which one is more important depends on the source impedance seen by the operational amplifier. In the case of the final topology from figure 8, the source impedance of U1 is mostly determined by the output conductance of Q1. For most JFET types with low voltage noise, the output conductance is in the range of 10–100 μS at 1 mA drain current and at V_{DS} equal to twice the $V_{GS(off)}$ magnitude. As output conductance scales linearly with drain current, a typical configuration with a total drain bias of 5 mA shows a resulting operational amplifier source impedance of 2–20 kΩ. There are numerous monolithic amplifiers available which provide very good noise performance for such medium source impedances; a typical choice would be the NE5534A.

A secondary noise source which so far has not found particular attention is the voltage reference V1. As equations 13 and 14 show, it should have low impedance and low voltage noise. Unlike the voltage reference needed within the drain load constant current source, the required DC voltage is just 2–3 V. Thus a few forward biased small signal diodes or light emitting diodes (LEDs) are suitable for a typical implementation. Initial measurements of 1N4148 parts suggest that at 1 mA bias current, the voltage noise per diode is well below 1 nV/rtHz at frequencies as low as at least 100 Hz, and thus negligible in most implementations.

9 Frequency Response, Compensation and Input Impedance

At low frequencies, the frequency response of the preamplifier is determined by the AC coupling and servo action. At high frequencies, it is mostly a function of Miller compensation, i.e. the value of C2. The single-ended amplifier topology considered in this article employs current feedback, i.e. the non-inverting input (the source of Q1) has low open-loop input impedance. As a first-order approximation, the closed-loop –3 dB bandwidth is then given simply by the time constant of the feedback resistor (R3 in e.g. figure 8) and the compensation capacitor. This approximation however is only valid if the input impedance of the non-inverting input, given by the reciprocal of Q1 transconductance, is much lower than the total resistance of the feedback network (R3 in parallel with R4). As in this low noise application the feedback network resistance is kept very low, this assumption is not valid. Secondary poles, introduced mostly from operational amplifier U1, further complicate the situation. Practical experimentation, or at least simulation, is the most accurate way to determine the appropriate value for the Miller compensation capacitor.

For low source impedances, meeting the frequency response specifications from table 1 is easy, as bandwidths of several MHz are achievable. However, with high source impedances, the input capacitance of the preamplifier comes into play. The gate-drain junction capacitance of the input transistors are the major contributors with this respect. Consider **figure 9** (for simplicity shown with standard AC coupling and without DC servo); if we assume that U1 is an ideal operational amplifier, the drain of Q1 sees a pure virtual earth, i.e. there is no AC voltage swing at this node. This eliminates

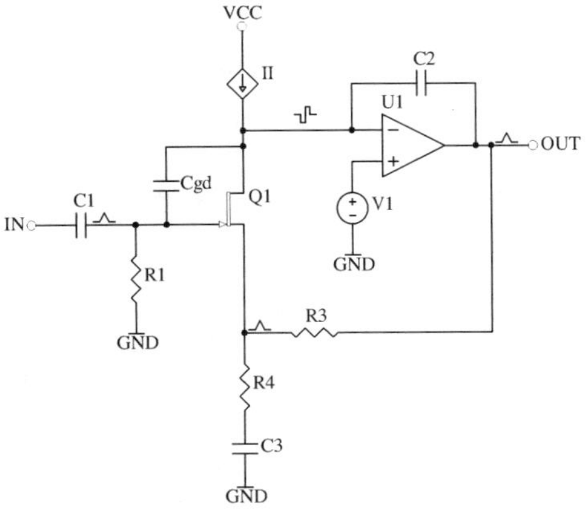

Figure 9: Input capacitance effects in the basic preamplifier architecture.

any Miller effect from C_{gd}, and the preamplifier input capacitance is simply equal to C_{gd}.

With real operational amplifiers, the node at the drain of Q1 is not a perfect virtual earth, but is subject to some voltage swing due to the finite open-loop gain of U1. As the operational amplifier open-loop gain basically resembles that of an integrator, the voltage at the inverting input must be a differentiated and inverted version of the output voltage. This is illustrated in figure 9 with the signal indicators at various relevant nodes. As long as the voltage swing at the drain of Q1 is smaller than the input signal, this is of no particular concern. However, once the drain voltage swing is similar or larger than the gate voltage swing, Miller effect will come into play. As the drain voltage is in a complex phase relation to the gate voltage, resulting Miller effect causes not just a straight multiplication of the input capacitance, but will also generate frequency dependent (i.e., because of falling operational amplifier open-loop gain, increasing with frequency) negative input capacitance terms. These interact with the source impedance and can result in gain peaking or even instability.

To eliminate these effects, the drain voltage swing would need to be smaller than the gate voltage swing within the full preamplifier bandwidth. This condition implies that the open-loop gain of the operational amplifier needs to be higher than the preamplifier gain. If the bandwidth of the preamplifier is 1 MHz and the gain as specified 60 dB, a gain-bandwith product in excess of 1 GHz is needed to meet this condition. While there are a few monolithic operational amplifiers available which at least approach such a large gain-bandwith specification, they inevitably have high power consumption, poor low frequency noise performance and operate just on ±5 V supply rails. This makes them incompatible with some of the specifications from table 1. Cascoding the input transistor Q1 is another possibility to reduce the drain voltage swing; however a cascode transistor inevitably has some noise contribution and reduces the available voltage headroom to implement the drain load. The most practical means to reduce the negative input capacitance is thus to just add a real shunt capacitor at the preamplifier input, which cancels the negative capacitance terms. Due to the frequency dependent nature of the negative input capacitance, the value of the added capacitance is necessarily a compromise. Higher values lead to better cancellation and thus less gain peaking at high frequencies, but imply higher positive input capacitance at low frequencies. Again practical experimentation, or at least simulation, is the easiest approach to determine the value which gives best frequency response flatness across the intended range of source impedances.

This completes the theoretical analysis and evolution of the basic preamplifier topology from figure 1. Section 10 now presents the actual low noise laboratory preamplifier design.

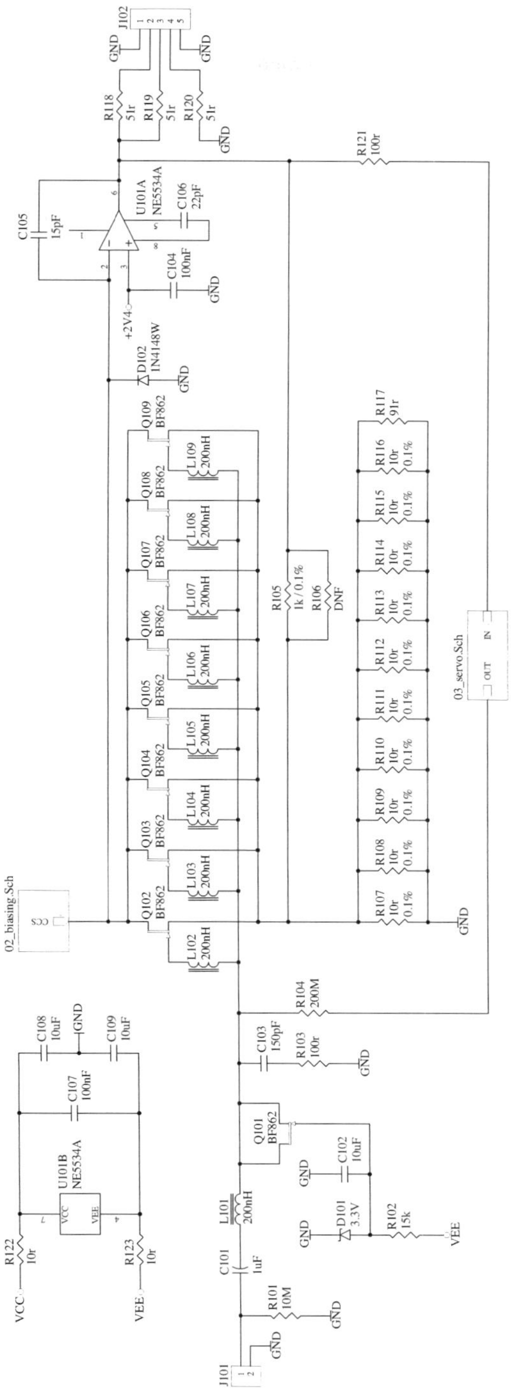

Figure 10: Main preamplifier schematic.

10 Actual Design

The schematics of the novel laboratory preamplifier are shown in figure 10–12. The circuit design closely follows the design strategies outlined in the previous sections. First, the BF862 JFET was selected as most suitable transistor type for the input stage because of its very low voltage noise, low input capacitance and good availability. After consideration of input capacitance effects and current noise, the number of paralleled transistors was set to eight. The measurement data presented in [16] suggested that a total drain current of 8 mA is sufficient to meet the preamplifier voltage noise specifications, while giving convenient headroom regarding quiescent current. The total transconductance of the input transistors is 88 mS, degeneration from the feedback network results in a net input stage transconductance of 81 mS. To reduce the chance of parasitic oscillation in the input JFETs, gate stopper inductors (L102–L109) were added. Ferrite beads with low Q are preferred for this application, as they show approximately resistive behaviour in the upper MHz region. This is most effective to dampen existing resonances (e.g. due to layout parasitics). Shunt capacitor C103 partially cancels the negative input capacitance of the preamplifier input stage. L101 aids the suppression of very high frequency interference and decreases sensitivity to electrostatic discharge (ESD). R103 dampens the resonance formed by L101 and C103. Diode-connected

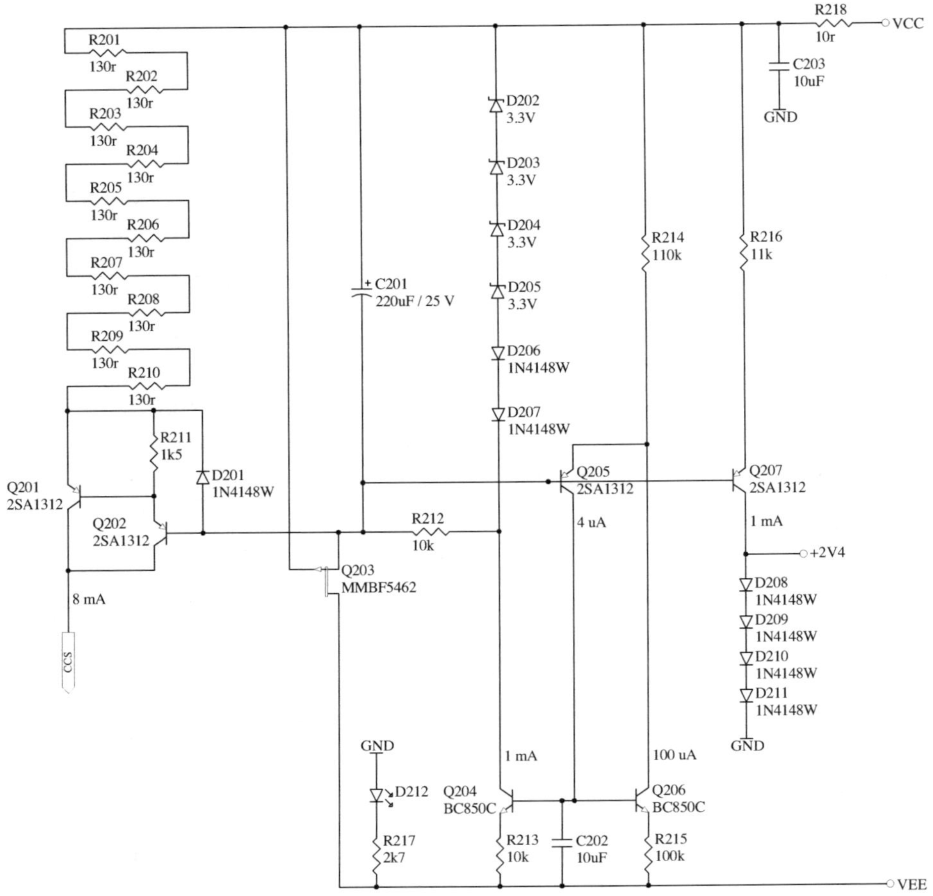

Figure 11: Low noise constant current source implementation.

Q101 and D101 clamp negative input voltages to about –4 V; positive input voltages are inherently limited as at some point the gate-source junctions of Q102–Q109 will become forward-biased. D102 provides further overload protection for U101. R101 is added to discharge C101 if the preamplifier is removed from a source with significant DC voltage.

The resistance of the feedback network is kept very low; the nominal values form a 1 Ω–1 kΩ divider. To achieve the desired gain accuracy of 0.5%, 0.1% resistors are used. 1 Ω resistors are not widely available in this precision class, so the value is synthesised from paralleled 10 Ω resistors (R107–R116). It was found that the measured gain of the preamplifier was consistently lower than predicted by the 1000:1 feedback ratio. While I have not been able to identify the exact reason for this behaviour, parasitic layout resistance which effectively appears in series with the 1 Ω shunt resistance is a likely cause. R117 compensates for this. By selecting this value and that of R106, the gain of the preamplifier may be set very precisely to 60 dB. However, measuring gain to an accuracy of much less than 1%

Figure 12: DC servo and JFET matching schematics.

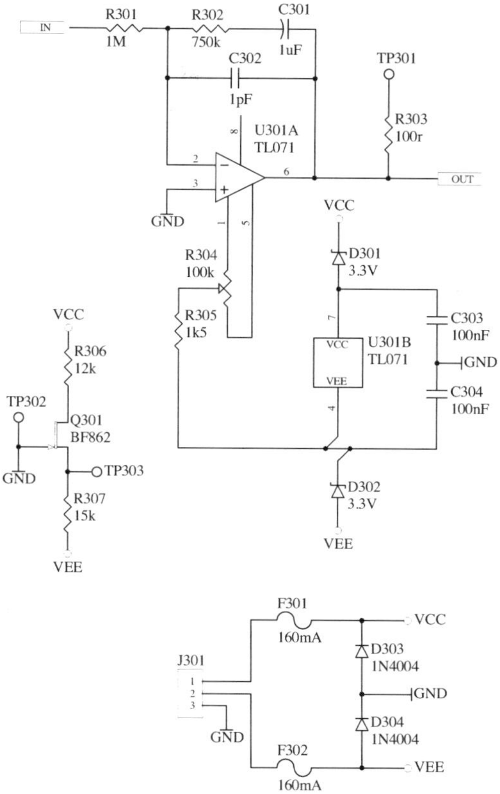

is difficult. In general, just using the nominal value of 91 Ω for R117 and not fitting R106 will be the most appropriate approach.

The constant current source which is used as input stage drain load employs Q201 and Q202 as pass transistor. The chosen part type has low noise (particularly a comparatively low excess noise corner frequency). D201 prevents reverse-bias (which could cause a significant noise performance decrease) of these transistors under overload conditions. The voltage reference which sets—together with the emitter resistors R201–R210—the output current of the current source, is formed by D202–D207. The use of several low voltage Zener diodes in series provides lower noise than one single diode with high breakdown voltage [1]; R212 and C201 further reduce noise by low pass filtering. As the Zener diodes are run at 1 mA rather than the nominal 5 mA, they each provide about 2.6 V DC. D206 and D207 extend this to a total of about 11.5 V. This is the highest DC voltage compatible with the minimum supply voltage and 5% Zener diode tolerance. To keep the current flowing through D202–D207 independent from power supply voltage, a bootstrapping arrangement is used [28]. It effectively consists of a 1:10 current mirror (Q204–Q206), and a 100 µA current source (R214) which is biased by D202–D207 and fed to the current mirror input. The output of the current mirror in turn is routed back to the D202–D207 diode string. Q203 provides fast and consistent startup; once the nominal bias conditions are reached, this transistor is pinched off and does not further affect performance. D208–D211 are used to form a 2.4 V reference voltage with low noise; it sets the drain-source voltage of the input transistors. D212 indicates power on.

U301 is the operational amplifier used for the DC servo. Its residual DC offset may be trimmed out with R304. D301 and D302 drop the supply voltage, as otherwise the absolute maximum rating of U301 would be exceeded at the highest specified preamplifier supply voltage of ±20 V. D303 and D304 protect the preamplifier supply if reversed power supply voltages were applied. F301 and F302 are resettable fuses which limit maximum input current under such conditions.

It is an unhappy fact that the gate-source voltage of JFETs shows considerable spread. For the BF862 at 1 mA drain current, I have measured values from −329 mV to −1.64 V from 25 samples, presumably

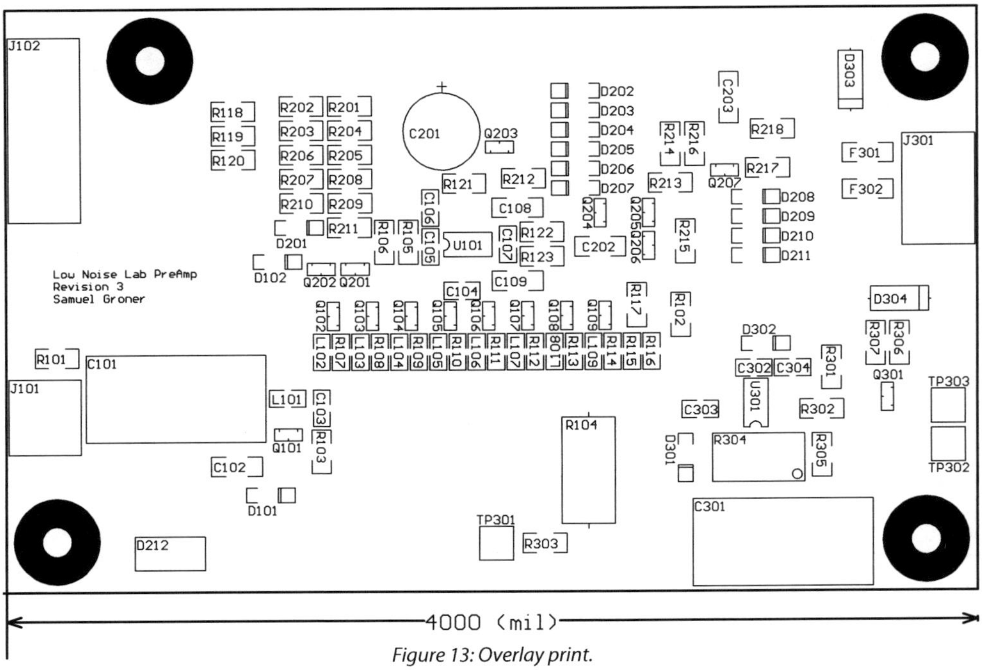

Figure 13: Overlay print.

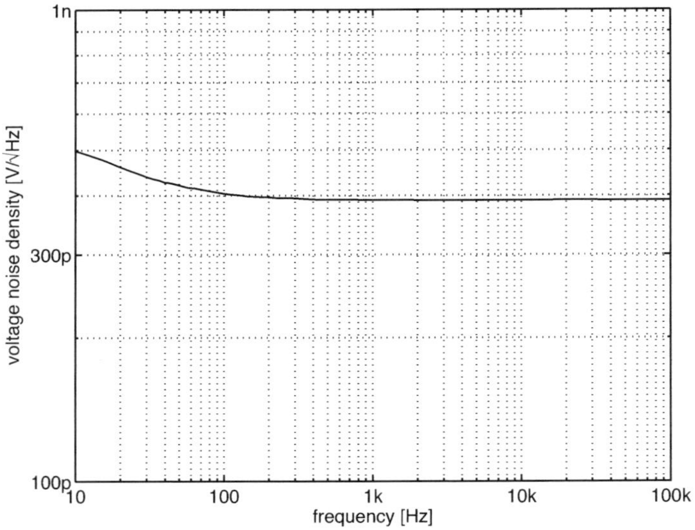

Figure 14: Equivalent input voltage noise density of the novel preamplifier.

all from the same batch. If transistors with such large parameter spread were combined in the input stage, the parts with high V_{GS} magnitude would carry far more drain current. Because of the JFET square law, this is detrimental to total transconductance, and thus noise performance.[6] Q301 is used to match the JFETs for the input stage; R306 and R307 set operating conditions similar to those in the actual input stage. Details on the matching procedure are found in section 12.

A four layer printed circuit board (PCB) was designed for this circuit. Significant effort was made to reduce the loop area of critical traces, mostly by extensive use of ground planes and careful placement. The resulting implementation is surprisingly insensitive to external magnetic fields, although at the very low noise levels achieved here it is still easily possible that e.g. a nearby power transformer ruins a measurement. The PCB overlay is shown in **figure 13.**

11 Performance Verification

Of prime interest for this preamplifier design is its voltage noise performance. **Figure 14** reveals a measured wideband figure of 390 pV/rtHz, and an extraordinary 1/f corner frequency of less than 10 Hz. These results are particularly notable as they are achieved at just 17 mA power consumption, and solely with the use of easily sourced standard components. The voltage noise level is very consistent, and stable just a few seconds after startup.

An isolated measurement of the constant current source drain load noise performance is depicted in **figure 15.** The thermal current noise of the used total emitter resistance of 1.3 kΩ is calculated as 3.56 pA/rtHz; above 100 Hz the constant current source implementation thus achieves a noise performance very close to the theoretical limit. Even at 10 Hz, the input referred voltage noise contri-

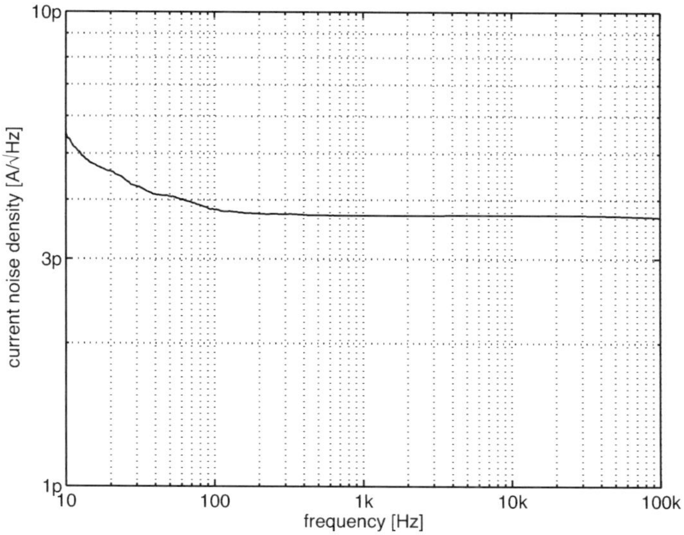

Figure 15: Output current noise density of the active drain load.

---

[6] Practical experimentation has shown that randomly selected input transistor combinations often show up to 20% higher voltage noise than matched specimen.

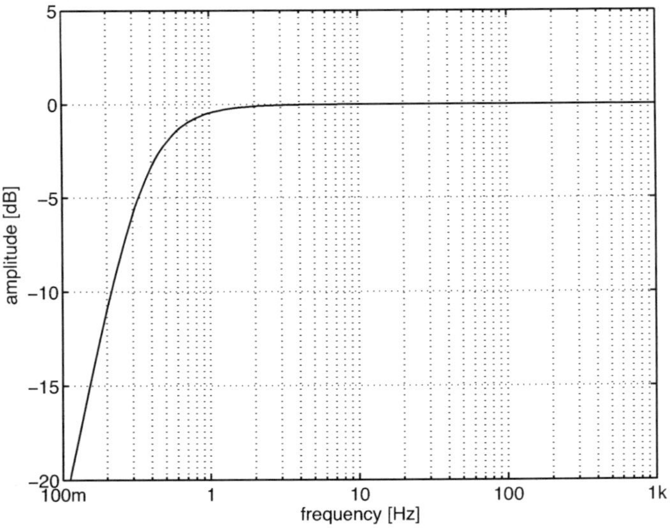

Figure 16: 100 mHz–1 kHz frequency response.

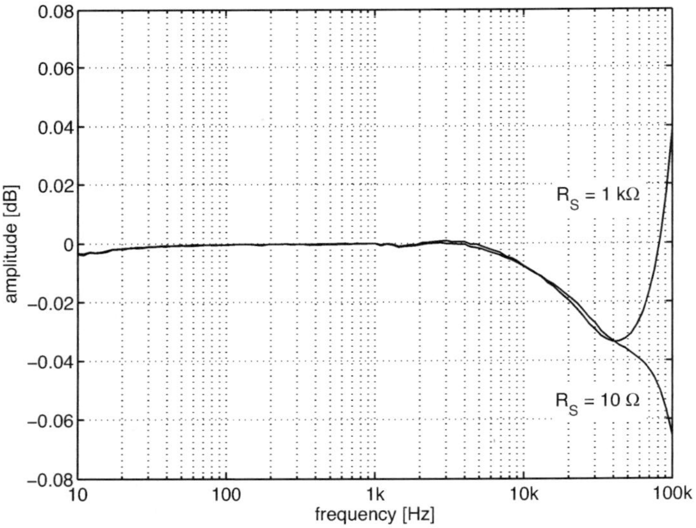

Figure 17: 10 Hz–100 kHz frequency response for two different source impedances.

bution of the drain load is just 68 pV/rtHz. This confirms the previous statement that the presented active drain load topology achieves exemplary noise performance.

Frequency response data is given in **figure 16 through 18**. Above 40 kHz, frequency response depends heavily on the source impedance. However, gain peaking due to negative input capacitance is relatively well controlled. Measuring 10 Hz–100 kHz frequency response to an accuracy of the frac-

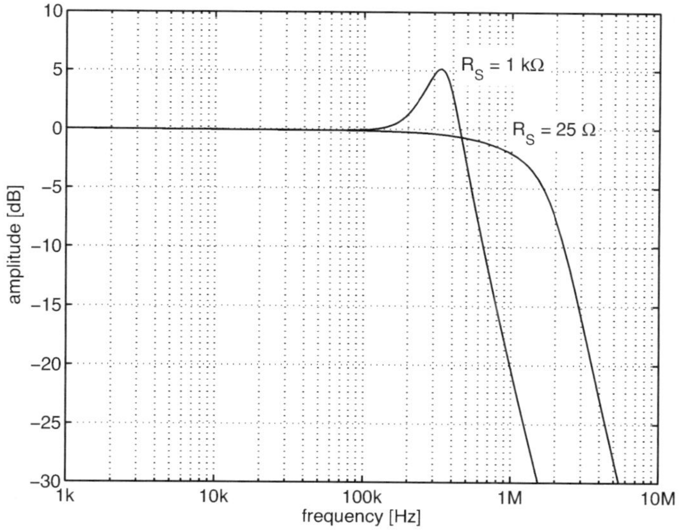

Figure 18: 1 kHz–10 MHz frequency response for two different source impedances.

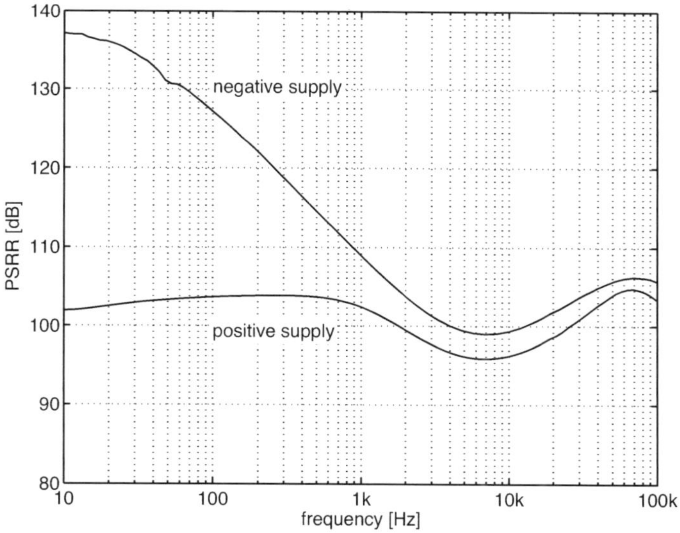

Figure 19: Power supply rejection.

tion of a dB requires exquisite care. While I believe that the presented results are made reasonably accurate by the use of special techniques which provide first-order cancellation of generator and analyser imperfections, verification with equipment specified to this accuracy is yet to be done. Input impedance was measured as about 250 kΩ at 1 kHz.

The power supply rejection of the preamplifier is shown in **figure 19**. While the positive power supply rejection is made sufficiently high through the use of an active drain load, the negative power sup-

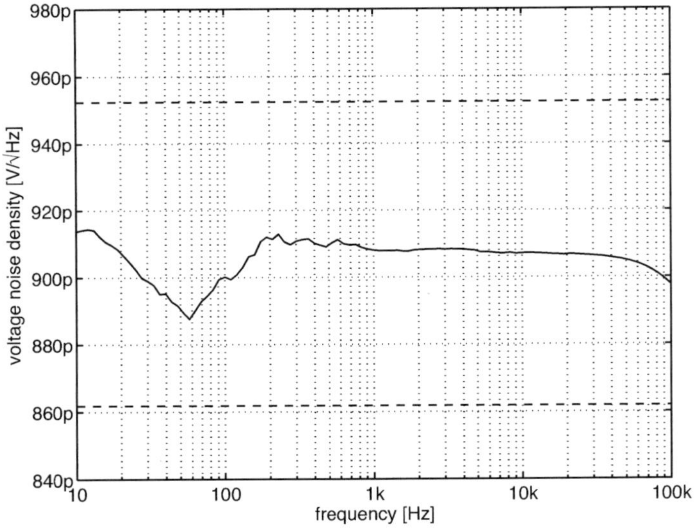

Figure 20: 50 Ω thermal noise measurement; dashed lines indicate 5% error bounds from calculated theoretical value.

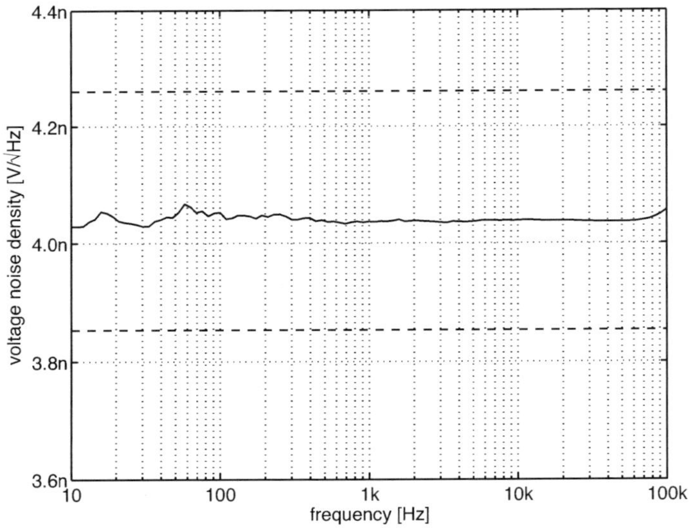

Figure 21: 1 kΩ thermal noise measurement; dashed lines indicate 5% error bounds from calculated theoretical value.

ply rejection (where there is no direct injection path in the input stage) is still considerably better at low frequencies.

As a final measure for overall preamplifier performance, it was used to measure the thermal noise of a 50 Ω and a 1 kΩ resistor. The results, along with 5% error bounds calculated from the theoretical values of 907 pV/rtHz and 4.06 nV/rtHz, are shown in figure **20 and 21**. Besides subtraction of the pre-

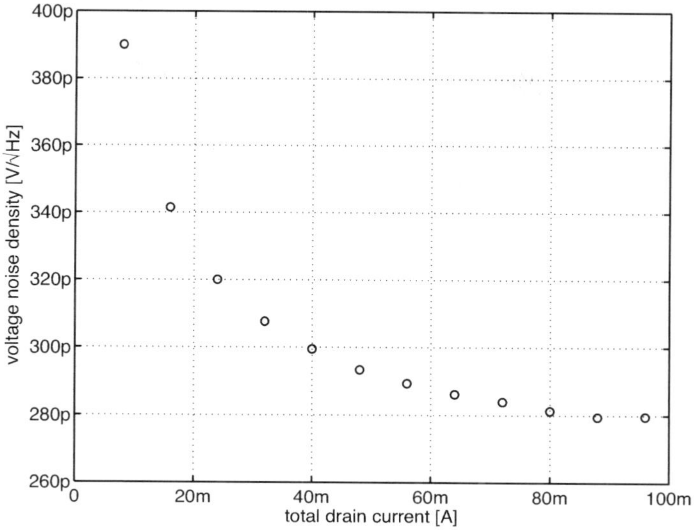

Figure 22: Equivalent input voltage noise at 10 kHz as function of total input stage drain current.

amplifier voltage noise and subsequent smoothing, the measurement data was not further manipulated and demonstrates the superb accuracy achievable with this novel design. The measurement of the 1 kΩ resistor also implicitly shows that preamplifier current noise is sufficiently low; due to its low magnitude, explicit measurement of equivalent input current noise was so far not successful.

The drain current of the input JFETs is kept relatively low in this design, to aid battery powering. How much voltage noise performance could be gained with higher bias? **Figure 22** plots the measurement results at 10 kHz, which were derived by adjusting R201–R210. The optimum drain current is 88 mA, with an impressive voltage noise density of just below 280 pV/rtHz. At 1 kHz, the same results are found. The high optimum drain current is indicative of the thorough minimisation of all secondary noise sources.

At lower frequencies however, the optimum drain current decreases as predicted in section 5. At 100 Hz, the optimum drain current drops to 32 mA, giving a voltage noise density of 310 pV/rtHz. With 96 mA, voltage noise performance worsens to 380 pV/rtHz. At 10 Hz, no improvement with drain currents above 8 mA was found; at 96 mA, the voltage noise density increased to almost 1 nV/rtHz. While I have not analysed the exact reasons for the noise behaviour at 10 Hz, I presume that it is dominated either from the constant current source and/or the NE5534A operational amplifier. Circuit changes as outlined in this article may provide some improvement in this respect. It should however also be kept in mind that two unmodified preamplifiers connected in parallel will likely provide even lower voltage noise (approximately 350 pV/rtHz at 10 Hz), at a total quiescent current of just 34 mA!

12 Construction Guidelines and Application Notes

Gerber files for this laboratory preamplifier, along with a detailed bill of materials, are made available at the authors webpage [29]; an overlay print layer is not included in the Gerber files, please refer to figure 13. The design is mostly based on surface mounted devices (SMDs), however no difficult to solder packages were used. Nonetheless, suitable tools (small tweezers and soldering iron tip) are a must.

It is best to start with the matching of the input transistors. For this, R306, R307, F301, F302, D303, D304 and J301 must be fitted, and the PCB connected to a ±15 V power supply through J301. Connect a volt meter between TP302 and TP303, set to the 10 V range. Now grab a JFET with tweezers, and place it at the position of Q301. Press it down gently with a small screwdriver, such that all three pins make contact to the PCB (no soldering is necessary). Now the voltmeter will read the V_{GS} of the transistor under test. Make a note of this measurement and continue with the next JFET. A V_{GS} matching of about 20% or better is desirable. Typically eight matching specimen are easily found within a batch of 25 transistors, and the procedure just takes a few minutes. Q101 does not need to be matched to the other transistors, of course.

To stuff the printed circuit board, it is convenient to first solder the two operational amplifiers. Then, proceed with the matched input transistors, and work from there towards the periphery of the board. The through hole components come last. Upon initial power up, verify that the voltage at TP301 is close to the measured V_{GS} of the matched JFETs (typically –500 mV), and the preamplifier output voltage within some mV near ground.

The single-ended input stage is somewhat sensitive to air flow, which can exist even within an enclosure. Although the resulting short-term drift signals are typically below 10 Hz, it is good practice to cover the sensitive parts (Q102–Q109, and to a lesser extent Q201 and Q202) with non-conductive foam (e.g. as sometimes supplied with IC samples). Cut a rectangular piece which covers slightly more than the area enclosed by C101, C201 and C301, and squeeze it such that C101 and C301/R304 hold it in place. With a shorted input and the preamplifier warmed up for a few minutes, you can now trim the output DC offset with R304.

Connection to a source should be carried out with a strictly coaxial signal path, up to the input connector J101. This minimises sensitivity to external magnetic fields, and prevents potential instability. To avoid damage to the preamplifier, its input should not be connected directly to a source with significant DC voltage and high peak output current capability. Either pre-charge the input (i.e. C101) through a 1 kΩ resistor for a few seconds, or power up the source with the preamplifier already connected. The rise time of typical power supplies is slow enough to limit the preamplifier input current to far less than 100 mA, which is uncritical. Similarly, if the preamplifier is removed from a source with significant DC voltage, the input AC coupling capacitor will remain charged for a few minutes. Shorting the input to ground during this time should be avoided.

13 Conclusion

In this article I have presented the design procedure for a low noise laboratory preamplifier. After establishment of the specifications, initial design considerations have led to the choice of an amplifier topology with single-ended JFET input stage. Detailed analysis of the various sources which contribute to the equivalent input referred voltage noise has shown that the choice of high drain current magnitudes for the input transistor pronounces secondary noise sources, mainly due to the square-law characteristics of the JFET. For certain circuit configurations, this implies that there exists an optimum drain current (below I_{DSS} which minimises total voltage noise referred to the input. Also it was found that paralleling transistors does not show this behaviour; thus it is a preferred method to reduce voltage noise.

To improve the power supply rejection of the preamplifier, a constant current source for the input transistor drain load was chosen. Practical design guidelines were presented which allow implementation of a suitable current source with very low output current noise. Further topological improvements have introduced a new low noise AC input coupling network which integrates a DC servo. The resulting filter response is of second order, which is beneficial for frequency response flatness, settling time after an input DC voltage step and protection against very low frequency signal content. Furthermore, the optimum choice of second stage operational amplifier type was discussed, and the presence and mitigation of negative input capacitance examined.

Measurements of the actual preamplifier revealed excellent performance particularly with respect to voltage noise, frequency response, power supply rejection and quiescent current consumption. The combination of these features makes the novel preamplifier design very well suited for a variety of practical laboratory measurement tasks.

Bibliography

[1] C. D. Motchenbacher, F. C. Fitchen: *Low-Noise Electronic Design,* John Wiley & Sons, 1973

[2] W. M. Leach, Jr.:*Fundamentals of low-noise analog circuit design,* Proceedings of the IEEE, volume 82, issue 10, pp. 1515–1538, October 1994

[3] Bruce Carter: *Op Amp Noise Theory and Applications,* 2008, available for download from http://www.ti.com/lit/ml/sloa082/sloa082.pdf

[4] Spencer Klein, Walter Innes, John C. Price: *Audiofrequency Measurement of JFET Noise vs. Temperature in a High Impedance Preamplifier,* Review of Scientific Instruments, volume 56, issue 10, pp. 1941–1945, October 1985

[5] G. V. Pallottino, Giuliano Vannaroni: *A Low-Noise Low-Input Conductance Preamplifier for Gravitational Research,* EEE Transactions on Instrumentation and Measurement, volume 34, issue 5, pp. 676–680, December 1985

[6] Steven R. Jefferts, F. L. Walls: *A very low-noise FET input amplifier,* Review of Scientific Instruments, volume 60, issue 6, pp. 1194–1196, June 1989

[7] Bruno Neri, Bruno Pellegrini, Roberto Saletti: *Ultra Low-Noise Preamplifier for Low-Frequency Noise Measurements in Electron Devices,* IEEE Transactions on Instrumentation and Measurement, volume 40, issue 1, pp. 2–6, February 1991

[8] J. Lepaisant, M. Lam Chok Sing, D. Bloyet: *Low-noise preamplifier with input and feedback transformers for low source resistance sensors,* Review of Scientific Instruments, volume 63, issue 3, pp. 2089–2094, March 1992

[9] C. Ciofi, M. De Marinis, B. Neri: *Ultralow-noise PC-Based measurement system for the characterization of the metallizations of integrated circuits,* IEEE Transactions on Instrumentation and Meas-

urement, volume 46, issue 4, pp. 789–793, August 1997

[10] Dennis Colin: *A Low-Noise Measurement Preamp,* Audio XPress, April 2007

[11] F. A. Levinzon: *Ultra-Low-Noise High-Input Impedance Amplifier for Low-Frequency Measurement Applications,* IEEE Transactions on Circuits and Systems I: Regular Papers, volume 55, issue 7, pp. 1815–1822, February 2008

[12] Charles Wenzel: *A Low Noise Amplifier for Phase Noise Measurements,* available for download from http://www.wenzel.com/pdffiles1/pdfs/lowamp.pdf

[13] Burkhard Vogel: *The Sound of Silence,* Springer, 2008

[14] A. van der Ziel: *Thermal Noise in Field-Effect Transistors,* Proceedings of the IRE, volume 50, issue 8, pp. 1808–1812, August 1962

[15] Ed Oxner: *Designing With Field-Effect Transistors,* 2nd edition, McGraw-Hill, 1990

[16] Scott Wurcer: *Low Noise Microphone Preamplifiers*, Linear Audio, Volume 1, April 2011 (part I) and Volume 3, April 2012 (part II)

[17] *Toshiba 2SK170 Data sheet,* November 2007

[18] Ashutosh V. Kotwal: *Characteristics of 2SK369 Junction Field Effect Transistor for Run 2 Calorimeter Preamplifiers,* March 1997, available for download from http://www-d0.fnal.gov/hardware/cal/notes/jfet_gm.pdf

[19] W. C. Bruncke: *Noise measurements in field-effect transistors,* Proceedings of the IEEE, volume 51, issue 2, pp. 378–379, February 1963

[20] F. A. Levinzon: *Measurement of low-frequency noise of modern low-noise junction field effect transistors,* IEEE Transactions on Instrumentation and Measurement, volume 54, issue 6, pp. 2427–2432, December 2005

[21] Samuel Groner: *A New Amplifier Topology With Push-Pull Transimpedance Stage,* Linear Audio, Volume 2, September 2011

[22] Y. Netzer: *The design of low-noise amplifiers,* Proceedings of the IEEE, volume 69, issue 6, pp. 728–741, June 1981

[23] *Vishay Beyschlag MMU 0102, MMA 0204, MMB 0207 Data sheet,* July 2006

[24] Frank Seifert: *Resistor Current Noise Measurements,* April 2009, available for download from https://dcc.ligo.org/public/0002/T0900200/001/current_noise.pdf

[25] *Vishay Precision Group HZ Series (Z-Foil) with Zero TCR Data sheet,* March 2010

[26] Walt Kester, Scott Wurcer, Chuck Kitchin: *High Impedance Sensors,* available for download from http://www.analog.com/static/imported-files/seminars_webcasts/371366216sscsect5.PDF

[27] Douglas Self: *Small Signal Audio Design,* Focal Press, 2010

[28] P. Williams: *DC voltage-reference circuits with minimum input-output differentials,* Proceedings of the IEEE, volume 57, issue 12, pp. 2180–2181, December 1969

[29] http://www.sg-acoustics.ch/analogue_audio/publications/

[30] Ovidiu Popa: *On the noise performance of low-noise input stages*, Linear Audio Volume 1, pp 147, April 2011.

The Way I see it...

Audio design is moving forward – or is it?

Stan Curtis

For as long as I've been interested in audio, and here we are talking of nearly 50 years, a constant source of pleasure, encouragement, amusement and controversy has been those magazine articles featuring new or improved circuit designs. Probably in my lifetime more column inches have been devoted to the discussion of audio than any other topic in electronic magazines and although the baton has now been passed to the internet (with a few honourable exceptions including "Linear Audio") the content is much the same. Or is it? Curiosity often drives me to try out many of the circuit ideas I see published yet sadly few make it to ongoing development. This is a disappointing outcome yet in many ways an outcome which is quite predictable. Why so? Well I think there are three groups which fit into this pattern. First many projects are misguided because they seek to overcome or minimize a problem that in truth is no problem at all. It is that well known characteristic of all engineers that no system is so good that it can't be improved and, of course, only their particular improvement will have any true merit. Now that is fine so far as it goes, particularly if the whole exercise is a somewhat academic exercise of the kind that leads to the "ultimate" supply regulator or what have you. But is it needed or does it have any audible effect upon the sound quality? Here we venture into controversial grounds because rarely has "perfect" measured performance correlated with perceived good sound quality or, perhaps more accurately, I should say that much of the time a clearly measured improvement in performance does not translate into a clearly identified improvement in sound quality. Indeed it is more often the case that "good enough" performance is certainly all that is needed to satisfy most listeners provided that a few essential criteria are met.

There are also far too many ideas out there being worked to death on the numerous internet forums. Everybody and his brother and sister's aunt is now publishing circuit ideas and a great many of them verge on nonsense leading me to wonder if they still teach Ohm's Law in schools today. And it is not the case that there has been an explosion of interest in DIY audio rather that everybody can now have a go following the advent of inexpensive, nay free, circuit simulation programs. Although I don't want to sound ancient it is the case that throughout my career designs have had to be worked up with pencil and paper before being tested through the use of a soldering iron and a bunch of components. And I have to say that the years of enduring the smell of burning resistors when first applying power to my prototypes certainly ensured that I finally cracked this audio design business. That is not to say I dislike simulations; I use them myself and they save a lot of legwork. But nothing replaces a solid understanding of circuit operation after which these programs can make design work a lot more productive. We've all got horror stories and I recall a young engineer showing me a convoluted circuit design that appeared to create a near enough perfect amplifier. Yet without knowing why every brain cell holding my years of practical knowledge was screaming "it will blow up". But no; he insisted on standing by his computer analysis and went off to build the prototype. As soon as power was applied the circuit went into viscous instability and the resistors started burning shortly after the driver transistors went to their maker. A post mortem revealed that his unique set of transistor approximations had managed to balance out some potentially nasty stability problems which the real life components failed to emulate. Nonetheless I still go ahead and build some of these circuits which have only ever existed in the memory of a laptop. And you know what? Sometimes I am pleasantly surprised by their performance although whether this is a testament to the designer's skill or the accuracy of modern simulations is a question that remains unresolved.

For my first contribution to "Linear Audio" I have therefore chosen not to offer some choice concoction of transistors but instead to focus upon something that is both fundamental yet quite ethereal. I would like you to think about why an audio stage of fairly mundane performance can sound so good to many independent listeners yet at the same time how can a seemingly minor circuit change having a measurable yet infinitely small effect also be heard.

Let's consider the first of these comments by reference to a recent change in my own household. This past Christmas I rewarded myself with a piece of sheer audio indulgence; namely a full size 24 track studio recorder using 2 inch tape. This particular beauty is a Lyrec and dates from the 1970s and is one of a generation of machines upon which some of the greatest music was recorded (**photo 1**). However it doesn't take long to appreciate why so many engineers welcomed the arrival of digital recording straight onto a hard drive. First there is the cost of tape which works out at about 12 euros per minute of recording time compared to disk space which is essentially free. Furthermore the use of such recorders is an acquired skill particularly when trying to time a smooth "drop-in" to synchronize a new recording with the tracks already laid down. And finally the measured performance or lack of, can be illustrated by just taking just two parameters; a signal-to-noise ratio of a little over 50 dB at a speed of 15 ips and channel to channel crosstalk which can be as low as 40 dB. So obviously such recorders have a sound quality to match their indifferent measured performance particularly when compared to the pristine perfection of a digital system. Well actually no, the subjective sound quality is to die for, meeting almost all the superlatives in the hi-fi reviewer's spell-checker. Now of course the Lyrec does have a reputation for good sound quality and its 24 audio boards are populated with the well regarded 5532 and 5534 chips. Yet my real favourite multi-track recorder; the 3M's M79 model does, to my ears, sound even better and yet it has audio boards populated with the early 741 chips. Yes 741s, those amplifiers of virtually nil slew rate and the all time winners of the "Worse Sounding Op Amp Award".

Lyrec

Given any choice I would always listen to master tape copies which used to be readily available from the record companies in the pre-pirating paranoia days but almost impossible to obtain these days. A few years ago I set up a demonstration where the same two-track masters were replayed from four tape machines ranging from a humble Revox A77 to a full size Studer A820. The sound quality was consistently stunning. Yes there were audible but minor differences between each machine due mainly to frequency response and channel separation; but I would have been happy to listen using any of these sources. By comparison I have spent many frustrating hours trying to optimize the playback of Compact Discs to give a sound I enjoy. On several recent occasions I've set up a transport to feed 6 or 7 different DAC designs and heard extraordinary differences in the way the sound is reproduced yet in measurement terms the differences were relatively minor.

Why should this be so? Well I put it down to a catch-all parameter which I call "coherence" which really describes a number of characteristics one of which is natural phase and timing relationships. The maintenance of the coherence of the music signal as it passes through various types of processing or format conversion seems critical to preserving the music in a form we all enjoy. Personally I find that you can take great liberties with the signal whilst it remains in the analogue domain and whilst the sound may change for the worse the overall performance is still enjoyable and acceptable. Part of this tolerance is down to the naturalness of distortion harmonics. Normally we measure harmonic distortion and inter-modulation distortion both of which are the result of a non-linear transfer function between the input and output. Thus it follows that distortion must be a bad thing. But to the listener the production of harmonics of the original sound is "mostly harmless" to quote Douglas Adams, as they simply add some harmonious colouration to the sound. Second Harmonic distortion adds some octave sound; Third Harmonic adds some octave plus the musical 5th above; the Fifth Harmonic adds some double octave plus the musical 3rd above; all being well until the Seventh Harmonic which adds some double octave plus what is approximately the musical Minor 7th above; a sound which may or may not be pleasing to your ear. Note that these are all musical relationships; a fact ignored by many engineers, and so most people find distortion not unpleasing and thus it is a not necessarily a bad thing. Above the Seventh Harmonic the musical relationships of the odd harmonics become less and less acceptable or even valid. Fortunately most audio circuits create harmonics which fall away rapidly in amplitude below the second and third but we are still left with the question of how little is little enough. Well here we can be guided by our tolerance of other audio equipment such as loudspeakers and analogue tape recorders where distortion levels of the order of 1% are quite the norm and give rise to little concern on the part of most listeners. My own tests over the years have shown the majority of listeners to be quite tolerant of 1% low order distortion and that no-one; absolutely no-one hears the difference between 0.1% and 0.01% distortion all other things being equal. Harold Leak's famous "Point One" philosophy still has some validity today.

So for my own part circuit design in the analogue domain is partly engineering and partly understanding the way listeners react to musical sounds. You could call it a blend of art and science. And it has served me very well over the years in terms of satisfied clients; magazines reviews and awards and, of course, numbers of products sold. Yet that same philosophy does horrify, not to say outrage, many other audio engineers who strive to produce the "perfect" amplifier or audio stage. Perfect that is in terms of the measured performance. Yet I have listened to so many such products and found them wanting; often seriously wanting, and I'm not so arrogant that I always know why. So I work with the recipes experience has given me to produce the best perceived sound for the manufacturing budget whilst always seeking ever better results.

However once we get into the digital domain this flexibility seems to break down because the music signal now seems to be a very fragile thing. For certain it seems that the conversion from analogue-to-digital and back again clearly has some way to go if my ears are not deceiving me. Could it be that if the conversion process reached some stage on the road to "perfection" then a digital copy of the recording would show that same resilience and tolerance as an original analogue recording? I suspect not because just comparing the effects of the pre-conversion processing and digital filtering demonstrates problem areas of which software algorithm obsessed design engineers seem oblivious. Of course there are some audio 'purists' out there who build CD players with absolutely no digital filtering and accept that there is a mirror image of the audio signal in the band above 22.05 kHz. The mathematics tell you that you risk horrendous inter-modulation distortion in the amplifier chain and tweeter as well as a mass of unwanted signal components scattered across the audio band courtesy of the real-life performance of the DACs. Well I have heard such players through wide-band "fast" amplifiers and there is no doubt that there is something to be said for the philosophy. The music seems to retain its vitality and flow, although for me there is something wrong about the sound during the louder passages. The simplicity of this approach has a certain emotional appeal but it has to be rejected because the problem of the Nyquist images frequencies is not solved; rather the listener goes into denial and pretends they aren't there.

The industry has embraced even more processing with up-sampling and up-conversion being seen as essential parts of the specification. This is in part a consequence of some inexpensive chipsets being developed for DVD players which have a PCM format with a maximum resolution of 24 bits/96kHz. To ensure compatibility this system has to permit the easy upwards conversion of 16 bit/48kHz and 16 bit/44.1 kHz signals. This can be a more than trivial task. When changing the sampling rate, it is easier to work with a multiple of the original signal's sample rate (*see Rick Lyons' article elsewhere in this issue – Ed.*). A two times over-sampling system will double the sampling rate, by adding an easy-to-calculate extra sample value in between each actual sample. However for an up-sampling filter to create a 96 kHz digital signal from a 44.1 kHz signal, some awkward mathematical calculations are necessary (the ratio being 2.1768707 or thereabouts) and you can find yourself wondering if one mathematician's filter algorithm becomes another man's "best guess". Certainly there is some approximation or "guessing" taking place because when I've fed the same digital audio signal to several processing chips and made the conversion to analogue using the same DACS; I can clearly hear the difference, indeed I can sometimes hear things that are just not right. And that just shouldn't be so.

I could go on for some pages on this topic but I guess my comments largely come down to this. The design of analogue circuitry is pretty much good enough and in my opinion often far better than is needed to satisfy the needs of the listeners; if the designer knows what he is doing. So we are developing new circuits simply because we can. If we can do better than why not and I for one will always enjoy reading about such new ideas. By contrast I firmly believe that the design of digital audio processing and conversion still has areas of weakness simply because it is so easy to lose that tenuous musical "coherence" I mentioned earlier. Yes we are restricted by the availability of processing chips from the big manufacturers but they can be used in ways unforeseen by those guys. It all comes down to our ingenuity and to judge from these pages there is plenty of that left to go round.

Book review – Valve Amplifiers (4th Edition)

Stuart Yaniger

Valve Amplifiers – Morgan Jones. Newnes (Oxford, UK), 2012. 4th Edition, Softcover 686 pages, ISBN 978-0-08-096640-3.

There are really two questions that a review of the latest edition of a longstanding and popular textbook must address. First, is the textbook worthwhile in general? Second, is it substantially enough different from earlier editions that it's worth the money to upgrade to the latest version? The former will be dealt with briefly, the latter more extensively. By way of full disclosure, I must give a bit of personal history. Some years ago on a business trip to London, I had arranged to meet up with a few of the characters I dealt with as a moderator on diyAudio.com (the largest web forum devoted to audio design and construction). One of them was a fellow who used the handle EC8010; he seemed extremely knowledgeable, straightforward, and very precise in his writing, a refreshing departure from many Net users; I was given to understand that his actual name was Morgan Jones, that he was an author, and that his books on tube design for audio were quite good. I was only vaguely interested in that- after all, I had a bookshelf full of Norman Crowhurst, Langford-Smith, Terman, the Radio Amateur's Handbook... We met up at Piccadilly Circus, where he turned out to be (at the time) a 40-ish fellow wearing faded jeans and a Wallace and Gromit t-shirt. My first words to him were, "I see we patronize the same tailor." Following several (perhaps more than several) beers and intense discussion about how tube amps ought to be engineered, I promised myself to get a copy of his book ("Valve Amplifiers," 3rd Edition), thinking it would be mildly interesting and wanting to support in some small way a smart, sane, and sincerely nice fellow- there's a shortage of these qualities in the world today. My assumption was that it would be a very nice book, but really add nothing to my library in the way of new information or understanding. I was quite wrong.

When the book arrived from Amazon, I started in on it and spent the next several days neglecting my family, occasionally slapping myself on the head, occasionally chuckling, and more than occasionally scribbling notes on the margins. Morgan took a very practical yet comprehensive approach to the design of valve amps and preamps. He was not afraid to use solid state components where they did their best- as voltage and current sources and references. Although I had designed and built quite a few tube amps over the years and had a pretty good grasp of the subject, every chapter had some nuggets in it that either I didn't know or systematized previously unconnected things I did know. And it was really well-written; unfortunately, there are superb engineers who cannot write and superb writers who do not understand engineering, but Morgan appeared to have that rare combination of talents.

This began a long and spirited correspondence between us, and we've managed to get together on occasion, so as part of disclosure I must admit that I've drunk his Scotch.

With that behind us, the first question can be answered quickly- there is a good reason that this book has become widely accepted and gone through four editions. I own perhaps 200 books on tube electronics. Exactly three of them are tattered, with broken spines, stained and solder-splashed pages, and copious notes on nearly every page. One of those three is the 3rd Edition of "Valve Amplifiers." What I didn't find in "Valve Amplifiers" was purple prose about intertransient silences, veils being lifted, or comparisons of soundstage between different vintages of Valvo EC91s . What I did find was straightforward exposition of fundamental design practices and data to back it up. The basic theme is, if you want to do something as irrational as design and build a tube amplifier in the 21st century, then for goodness sake, engineer it properly. Enough said, if you don't have the 3rd Edition, stop reading this and order the 4th Edition today.

The next question is the value of upgrading from Version 3.0 to Version 4.0. Much has changed since the appearance of the 3rd edition- the widespread availability of accurate and easy-to-use simulation software such as Duncan Munro's PSUDII power supply calculator, the appearance of low cost high voltage depletion mode MOSFETs, new and better voltage references, and perhaps most importantly, Morgan's recent career-switch from teaching and broadcast engineering to the development of low noise high precision sensors and electronics. His new specialty informs many of the changes in the 4th edition.

The 4th Edition is somewhat longer (about 100 pages more) than the 3rd. Besides several new topics and new projects, much of the remaining material has been rewritten, with fresh insights marbling the meat. As before, Chapter 1 starts with a review of basic electronics; in reality, if you approach this book without fully understanding the basics of Ohm's Law, Kirchhoff's Law, the concepts of reactance, and a basic understanding of the function of semiconductors, the sledding will be rough. But at least the ground is covered, however quickly. An important part of this chapter is the derivation of basic feedback equations. A flavor of Morgan's engineering approach (and one of those little nuggets that makes this book so worthwhile, as he hacks the reader's way through the underbrush of audio mythology) is this observation:

In the late1970s, when cheap gain became readily available, designers became very excited by the possibilities and implications of the feedback equation, and set out to exploit it by designing amplifiers that were thought to have high levels of feedback. In practice, these amplifiers did not *have high levels of feedback at all frequencies and power levels, and it was the* lack *of feedback to linearise these fundamentally flawed circuits that caused their poor sound quality.*

There's also considerably more information added on contemporary LEDs and other voltage references. I would want to see the Ebers-Moll equation introduced earlier, and perhaps some simple diagrams of semiconductor junctions would clarify some of the text, but these are relatively minor points.

Chapter 2 is where the non-specialist will get a taste of how tube circuits work. Morgan covers the basic building blocks of more complex circuits, including common-cathode voltage amplifiers, cathode follower buffers, mu followers, cascodes, and the like. His treatment is not as deep as one will find in some of the classic texts (he specifically avoids the use of complex variables and calculus; in his words, "*All that is left is ordinary algebra, and although there are lots of equations, they are timid, miserable creatures, and quite defenceless.*"), but it will give the reader an understanding of how those circuits work and the basic design equations. Most of this chapter is unchanged from the 3rd Edition, but the use of new depletion mode MOSFETs to construct simple, high performance current sinks/sources is new and a valuable addition.

The book then continues on to a discussion of distortion and noise, one of the meatiest chapters. The best parts of the 3rd Edition have been retained, but as a consequence of the professional work in which Morgan has been engaged, there is a great deal more information on modern LEDs for biasing, noise optimization, current noise, and noise from voltage references. There are also boatloads of data presented to support the theoretical considerations. Much of the data is surprising, and use of it is made later in the book. The additional material here alone is, in my opinion, worth the cost of the new edition, especially if the reader wishes to design (or understand the design) of preamps for high source impedance transducers like condenser microphones.

Other new material here includes discussion and measurement of several distortion-cancelling schemes, including the Harmonic Equalizer, a neat tweak for push-pull circuits originally patented by Western Electric. One other noteworthy addition to this chapter is a discussion of the now-canonical Baxandall distortion-versus-feedback analysis; what sets this apart from other treatments is a critical examination of the limitations of Baxandall's simple model and where it may not be applicable.

As we know too well, real electronic components are not quite the idealized bits that we plug into design equations and SPICE models. Whereas the usual audiophile tendency is to wax mystical and poetical about different types and brands of components, Morgan takes a more analytical approach and examines what the physical and electrical differences are, why those differences exist, and how to account for them when designing circuits and choosing parts. Since the last edition of the book, he has done extensive measurements; refreshingly, he has not followed the usual audiophile practice of setting up test circuits which deliberately torture components to find differences that will not affect the likely circuits in which the components are used. Nor does he try to parse out conclusions from insanely small differences (e.g., resistor nonlinearities at -150dB) that could just as easily result from test equipment. Instead, a straightforward approach is taken, looking at construction of components, how that affects the measured performance, the relevance of that performance to real-world audio circuits, and how the designer can make rational choices for components in different circuit positions.

One of the more novel aspects presented in the 4th Edition is a new model for capacitors. This involves a large imaginary capacitance (i.e., a real term in the impedance that is inversely proportional to frequency) in series with a more typical capacitor model. Unfortunately, his deliberate choice to avoid complex numbers makes this slightly more opaque and there's seemingly no connection to intuitive physical factors, but nonetheless, this fits measured data quite well. Interestingly, the imaginary capacitance is grossly different between otherwise similar dielectrics; for example, there's a surprisingly large difference between polypropylene and polystyrene (two excellent capacitor dielectrics). A possible figure of merit for choosing capacitors is a large imaginary capacitance that has relatively little variation with frequency. Not surprisingly, Teflon wins that contest. I would be very interested to see further measurements and analysis of this interesting model, and some sort of correlation to well-understood physical factors.

Much new material appears in the chapter on power supplies, thanks to the emergence of PSUDII simulation software and better measurement capabilities. The comparison among different rectifier types (silicon, hard vacuum, mercury vapor) has been expanded, but the conclusions remain the same- avoid mercury vapor unless you particularly like the nice purple glow, the switching noise, and the short wavelength ultraviolet light. And as in earlier editions, Morgan presents some interesting tube voltage regulator designs but clearly believes that, for constant voltages or constant currents, solid state is a better path.

Shunt regulators have come into vogue in the past few years. Their inefficiency is often compensated for by low noise, so they are especially popular for powering cascodes and similar circuits with poor power supply rejection. Most of these circuits make use of a series resistor and a feedback-controlled pass device in shunt. In the novel Statistical Regulator, Morgan takes advantage of the fact that for two identical voltage references in series with uncorrelated noise, there is a 6dB increase in voltage drop, but only a 3dB increase in noise. Combining this idea with the simplicity and high performance of the depletion mode MOSFET cascode constant current source, a regulator design with cheap parts, utter simplicity, no feedback or adjustments, and very low noise can be designed and built. Measurements of all critical bits of the circuit are shown and explained in detail- as with most of the designs in this book, this one is intended to equip the reader with the information and understanding to allow him to design variations of the basic circuit for different applications rather than just copying part for part.

In the power amplifiers chapter, the analysis of three classic circuits (the Williamson, the Quad II, and the Mullard 5-20) is retained from earlier editions. These analyses reinforce the earlier lessons about feedback and how functional blocks interact. The Crystal Palace "dream amp" design from the 3rd Edition, a medium power class A design exercise, has also been retained, albeit with some interesting improvements like differential feedback. The CP is an excellent pedagogical vehicle, and having spent a few evenings with this amp, I can attest to the performance. But much of the complication was unnecessary, the requirements for the output stage can be daunting, and many of the functional blocks (particularly the cathode followers driving the output stage) can be done better with newer devices. Thus was born the Bulwer-Lytton, a low-power updated version of the Crystal Palace using paralleled 6S4 triodes as the output tubes. The new design exercises in the Bulwer-Lytton include using the Harmonic Equalizer in the output stage and the design of MOSFET source followers and their performance versus cathode followers. Interestingly, Morgan stops short of showing the use of source followers to take output stages (especially triode) into class AB2 for higher power and efficiency. Perhaps in the 5th Edition?

The book concludes with a preamplifier chapter that has the same thorough analysis of noise as in the 3rd Edition, but much added material on volume and balance controls, and a detailed examination of the Baxandall tone controls. New data is presented on the performance of pentodes in feedback line amps. As well, there's a new balanced hybrid phono stage, with a differential cascode input using (gasp!) bipolar transistors for optimum noise performance. RIAA design is without question difficult, and if nothing else, the reader will come away with a new appreciation for just how difficult it is and an understanding of why most designs have mediocre performance.

To this point, I've outlined the many virtues of this book. Now we need to take out the sharp knives: what are the shortcomings? Though quite comprehensive, "Valve Amplifiers" 4th Edition does not cover everything, though it does pack quite a bit into slightly under 700 pages. If you want detailed treatment of directly heated triodes or single ended amps, there's slim pickings (though the Scrapbox Challenge is a nicely executed single ended amp). The treatment of transformers is very good, but not as complete as that in Menno Vanderveen's "High End Valve Amplifiers 2" or the classic "Radiotron Designer's Handbook." Likewise, if one wants a broad treatment of phase splitter and output stage topologies, with worked examples, Crowhurst's "Understanding Hifi Circuits" covers that ground very well- by contrast, Morgan chooses the topologies which he thinks are the best and most useful and examines them in far greater detail. Menno's treatment of output stage and exotic transformer topologies is also more complete than the coverage in Valve Amplifiers. Bootstrapping, Schade feedback schemes, nested feedback, current source amplifiers, and screen drive are left mostly unmentioned. Musical instrument applications are also relatively untouched. You also won't find tubed DACs, I-to-V converters, tuners, or microphone preamps, though in fairness, the tools for the reader to understand the requirements of these applications and do a creditable design are certainly there. The biggest fundamental omission is a consequence of the decision to eschew complex numbers; thus, any quantitative explanation of amplifier stability is compromised, and the reader will not see complex planes and pole-zero analyses. I would also like to see design problems at the end of each chapter- I'm tempted to write that myself as a supplement to the book. After seeing explanations and examples, nothing cements understanding more firmly than working through numerous problems.

So, where does this book fit? Unlike some other recent books on tubes, the English is clear and elegant, and at times quite amusing. The pedagogical treatments are clear and enjoyable, and unlike most classic texts, it's a book that can be read for pleasure. The value here is the pedagogy- Morgan is the professor you always wanted but never had. Other books cover specialty topics more extensively, and this is not a step-by-step "how to" as much as it's an effort to make the reader understand "why." But "Valve Amplifiers" is a dazzling toolbox which will equip the reader to analyze and design tube circuits, no matter what the application. After owning earlier editions and now having absorbed the new material in this one, I can't imagine how I lived without it.

Book review - The Sound of Silence (2nd Edition)

Ovidiu Popa

The Sound of Silence – Burkhard Vogel. Springer Verlag, (Heidelberg, Germany), 2011. 2nd Edition, Hardcover, bound, 752 pages, ISBN 978-3-642-19773-4.

Writing a book about low noise electronics is a difficult task today; a great deal of information was published in the last 50 years, and quite a few definitive resources already exist, both at the academic and a more practical level [1] [2].

When the first edition of Burkhard Vogel's "The Sound of Silence" was published in 2008 it was obviously not just another book on an already beaten to death topic. While using a level of details that would make happy most of the academia audience, the book was also very practically oriented, presenting a number of detailed options for implementing a RIAA preamp, complete with schematics and construction details. In fact, exactly that combination of theory and practice made this book remarkable. While the result-oriented hobbyist could simply extract the schematics and follow up an implementation path, the inquiry minds could also further dig in the text and find out the theory and experimental work behind each and every design decision.

I was really curious about the books' 2011 2nd edition; what could it bring extra to the table? After all, three years is not that long of a time. Noise theory didn't change in the last fifty years, and the principles behind a good RIAA preamp also didn't change.

The 2011 2nd Edition revealed to be a journey both in the future as well as in the past. While a modern analytic method of evaluating the noise performance was added (using Mathcad) in Ch. 12, no less than two full chapters were added on the noise theory (Ch. 7) and practice (Ch. 17) of vacuum triode driven phono preamps. Strictly from a noise perspective, there's no reason why a designer should choose tubes over solid state devices. However, tubes are the devices of choice of many audiophiles and, before any attempt is undertaken to analyze the pros and cons of vacuum state vs. solid state devices and any particular implementations, this preference is to be acknowledged. From this perspective, a new resource on tubes noise, adapted to the 2011 tools and design methodology is to be appreciated.

Another couple of new chapters (24 and 25) are dedicated to a problem often ignored in audio systems design – hum. Hum is a source of many frustrations among audio designers, DIY communities and audiophiles, after all what's the point of having an amazingly low noise figure of the signal processing channel if the overall system is plagued by a stubborn mains frequency (and harmonics) hum in the speakers? Following the systems noise characterization methodology, a "Hum Figure" is introduced and used to analyze and characterize phono preamps.

Not strictly noise related, but nonetheless very useful, is Ch. 3. It describes a few basic details about vinyl recording technology and replay cartridges. This chapter could very well expand or even be the subject of an entire new book – I consider this at the very foundation of building a good quality RIAA preamp.

Actually the entire Part II – Noise Basics was revisited in the 2nd edition. While certainly not getting into the intricacies of the physics foundation of noise, this part is a very concise and complete view on noise mechanisms in electron devices (from resistors to bipolar transistors and JFETs, along with the already mentioned tubes). The next logical step is to analyze the two ends of the audio chain, the gain blocks (operational amplifier architectures, instrumentation amplifier architectures) and the sources (cartridges and input transformers). By its scope and structure, the entire Part II is the closest to the educational goal of this book: in a simple and attractive form, it lays down the foundation of noise calculation and it clearly illustrates the engineering concepts and methods of analyzing and designing low noise amplifiers. For those less skilled into electronics engineering it reveals some counter-intuitive facts (like differential gain also means additive noise, so the benefits of differential gain stage architecture come to a 3dB noise penalty). But, after all, how much signal to noise ratio do we really need for playing vinyls? Ch. 11 has the answer, and a few examples are illustrating the theoretical calculations.

Chapter 19 is a deep dive in the land of RIAA networks. Both passive and (fully or partial) active, balanced and unbalanced, in the ideal and real (considering the finite input and output impedances/admittances) cases. This chapter is a good example of the holistic approach in this book: the focus is not on the frequency response calculations, but on the integration considerations. Understanding such potential integration issues is the key for a successful RIAA amplifier design and as an example, passive RIAA correction networks have a considerable impact, lowering the RIAA amplifier overall headroom, hence such networks are recommended mostly for tubes implementations.

A very important topic, measurements of noise and frequency response, is covered in chapters 22 and 23. The concept of a "measuring amp" is introduced, as a precision gain block, followed by a precision rectifier, logarithmic converter and filter. The filter design options are covered in chapter 23, together with the anti RIAA configuration, very useful in evaluating the overall RIAA precision. Several common measurements pitfalls (like a missing RBW specification in FFT analysis and incomplete S/N spec-

ification) are addressed and explained.
Finally, the part that probably is the delight for DIYers, chapters 27 to 29, are dedicated to a very practical approach of the RIAA phono amp. A flexible modular phono amp configuration is defined, starting with the system requirements and going into the ultimate design details (power supply, soft start, and wiring). Calculated and measured performances are shown in chapter 29. These chapters are concluding the rich content of this reference book.
What is this book missing? I would say nothing. I am not aware of any other book with a better coverage of the phono amp analysis, design and implementation. What one could think about this book is giving out to much! While certainly there's nothing useless in its rich content, perhaps the Mathcad sheets are overwhelming for the less knowledgeable. Besides, although a 30 days evaluation is available, Mathcad is not free software and, with a price tag well over $1,000 for a full license it's barely accessible to the non-professionals. On the other side, there are sometimes too many low level details that may obscure the big picture, in particular for the more skilled engineers. Fact is, while the book is trying to address the needs and make everybody happy, there is a risk of actually frustrating everybody. I may suggest for a future edition reorganizing the content in a more logical way, such as a complete journey, from system requirements, to architecture design, building blocks, part selection (which certainly needs a 2011 and beyond level update, anyway), auxiliary circuitry (power supply, etc...), performance analysis, performance measurements. All low level details could then be placed in Appendixes, intended for those looking after a deep dive. Such an approach would certainly define and better focus on a target audience.
In conclusion, this is a reference book that shouldn't be missing from anyone's library, be it the low noise design professional, or the enthusiastic audio DIYer. Anybody with an interest in low noise audio amplifiers, from old fashion tubes aficionados to the latest solid state low noise designers, will find this book useful. It is also a very pleasant lecture and, once the reader gets accustomed to the content layout, can also serve as an excellent quick reference. Thank you Mr. Vogel for putting together and maintaining this book!

References:

[1] Albert Van der Ziel, "Noise in Solid State Devices and Circuits", Wiley-Interscience; 1st edition (May 13, 1986), ISBN: 0471832340
[2] C. D. Motchenbacher, J. A. Connelly, "Low-Noise Electronic System Design", Wiley-Interscience; 1st edition (June 29, 1993), ISBN: 0471577421

Made in the USA
Monee, IL
21 September 2026

63875502R00105